BEHAVIOR AND ITS CAUSES

STUDIES IN COGNITIVE SYSTEMS

VOLUME 16

The titles published in this series are listed at the end of this volume.

BEHAVIOR AND ITS CAUSES

Philosophical Foundations of Operant Psychology

by

TERRY L. SMITH
University of the District of Columbia, Washington, DC

KLUWER ACADEMIC PUBLISHERS
DORDRECHT / BOSTON / LONDON

A C.I.P. Catalogue record for this book is available from the Library of Congress.

ISBN 0-7923-2815-9

Published by Kluwer Academic Publishers,
P.O. Box 17, 3300 AA Dordrecht, The Netherlands.

Kluwer Academic Publishers incorporates
the publishing programmes of
D. Reidel, Martinus Nijhoff, Dr W. Junk and MTP Press.

Sold and distributed in the U.S.A. and Canada
by Kluwer Academic Publishers,
101 Philip Drive, Norwell, MA 02061, U.S.A.

In all other countries, sold and distributed
by Kluwer Academic Publishers Group,
P.O. Box 322, 3300 AH Dordrecht, The Netherlands.

Printed on acid-free paper

Printed in the Netherlands

For Mom and Dad

Behaviorism, as we know it, will eventually die--not because it is a failure but because it is a success. As a critical philosophy of science, it will necessarily change as a science of behavior changes, and the current issues which define behaviorism may be wholly resolved. The basic question is the usefulness of mentalistic concepts. (Skinner, 1969, p. 267)

SERIES PREFACE

This series will include monographs and collections of studies devoted to the investigation and exploration of knowledge, information, and data-processing systems of all kinds, no matter whether human, (other) animal, or machine. Its scope is intended to span the full range of interests from classical problems in the philosophy of mind and philosophical psychology through issues in cognitive psychology and sociobiology (concerning the mental capabilities of other species) to ideas related to artificial intelligence and computer science. While primary emphasis will be placed upon theoretical, conceptual, and epistemological aspects of these problems and domains, empirical, experimental, and methodological studies will also appear from time to time.

While most philosophers and psychologists tend to believe that the rise of cognitive psychology has occurred concomitant with the decline of operant psychology, Terry L. Smith contends that nothing could be further from the truth. He maintains that operant psychology has discovered (and continues to discover) reasonably well-confirmed causal principles of intentional behavior, which go beyond what cognitive psychology can provide, while cognitive psychology, in turn, has the potential to supply analyses (and explanations) that account for them. Smith thus advances a surprising but nonetheless illuminating perspective for appreciating the place of operant conditioning within the discipline of psychology in this rich and fascinating work.

J. H. F.

TABLE OF CONTENTS

PREFACE

In the summer of 1980, I attended Professor Dudley Shapere's National Endowment for the Humanities Summer Seminar on the philosophy of science. The topic was scientific revolutions. Each participant was to choose a revolution and do a case study of it. Like many other philosophers, I thought psychology had undergone a revolution in the 1960's, so I chose that episode for my case study. I focused upon Noam Chomsky's (1959) critique of B. F. Skinner's (1957) *Verbal Behavior*.

When I outlined my preliminary analysis for Professor Shapere, it dealt with the issue of whether Chomsky's case against Skinner would have been decisive within a behaviorist epistemology. He listened quietly, nodding in agreement occasionally, then surprised me when I had finished by asking only this question: "What happened to the Skinnerians?"

In truth, I had never thought about it. I had (like just about everyone else) read Kuhn (1970), and so almost reflexively I had interpreted cognitive psychology and behavioral psychology as competing paradigms (see Leahey, 1992, for a discussion of how common, and mistaken, this interpretation is). Cognitive psychology clearly was on the rise, so I inferred that the Skinnerian program must be on the decline. Indeed, I thought it must have just about disappeared by now. Professor Shapere's question, however, implied that this outcome was not a foregone conclusion, so I took the prudent course and replied that I did not know what had happened to the Skinnerians. He suggested this might be worth looking into. That proved to be a fruitful suggestion. What I discovered was that during the 1960's, the Skinnerian program had actually *grown* at an *accelerating* rate. This baffled me. How could operant psychology have survived, and even prospered, in the midst of "the cognitive revolution"? My interest soon shifted from the narrow question of whether Chomsky had refuted behaviorism on its own terms to the broad puzzle of how the operant program had managed to grow (and grow rapidly) when cognitive psychology was also growing rapidly.

The following essay is my attempt to solve this puzzle. My approach is philosophical, rather than historical or sociological--i.e., my solution refers to the concepts and theories of the program itself, and not to the "considerations of authority and power" (Kuhn, 1992, p. 8) that constitute the major alternative to philosophical explanation in this domain. I start with a working assumption: Operant psychology is unlikely to have grown in the manner it did without having something important (and essentially correct) to say about behavior. I attempt to find out what this something is, and to explicate its relationship to behaviorism and to cognitive psychology. My major conclusions are: (a) that operant psychology has formulated (and continues to formulate) reasonably well confirmed causal principles of intentional behavior (including the intentional behavior of human beings), whereas cognitive psychology has not formulated (and cannot by its very nature formulate) such principles; and (b) that cognitive psychology has the potential to provide (and to some extent already does provide) an analysis of the processes that underlie (and therefore partially explain) the principles mentioned in (a), whereas operant psychology has failed (and seems likely to continue to fail) to do so--even though Skinner based his philosophy on the thesis that the operant program was destined eventually to succeed in doing so. In other words, operant psychology and cognitive psychology complement one another.

This carries the account about as far as a philosopher can go. What I show is simply that the two programs fit together--or more specifically, that not only do their major assertions not contradict one another, but the explanatory strengths of one are the explanatory weaknesses of the other. I would suggest (although I must leave it to historians to test the idea) that the actual concurrence of the growth of the two programs is due to the fact that they are different aspects of a single process of historical development.

The following analysis is meant to address the concerns of philosophers, cognitive psychologists, and operant psychologists. In order to say something of interest to what is, in the world of philosophical monographs, a rather diverse audience, one must make some decisions about when to address whom about what. In general, questions are raised in the order they would arise for a philosopher who is skeptical about the relevance of operant psychology to any significant question about the intentional behavior of a normal adult human being. Actually, this seems to be the opinion held by most professional philosophers, whether they be specialists in the philosophy of psychology or generalists who touch upon Skinnerian psychology only occasionally--perhaps in teaching an introductory philosophy course. It is likely that more than a few cognitive psychologists hold this opinion as well.

These two groups--philosophers and cognitive psychologists--are the primary audience. The secondary audience consists of operant psychologists. I assume the primary audience to be familiar with the basic concepts and principles of cognitive psychology. I do not, however, assume familiarity with the basic concepts and principles of operant psychology. Indeed, I assume the opposite. Therefore, Part I spends a significant amount of time going over ground that will be familiar to operant psychologists. Part II draws some of the philosophical implications of the concepts and principles discussed in Part I. Some, but not all, of this material will be familiar to operant psychologists. I believe they will find it to be consistent with their understanding of their discipline.

In Part III, however, the interpretation of operant concepts and principles becomes more open to challenge by operant psychologists themselves. The objective is to explicate the relationship between radical behaviorism and Skinner's own scientific research. Little progress can be made on this front, however, without using the term *radical behaviorism* in a manner that is consistent with Skinner's intent. Unfortunately, there is a widespread inclination even among operant psychologists to tolerate a much looser usage than Skinner's. I therefore try to reconstruct what Skinner meant by his coinage. I then use this reconstruction to make sense of Skinner's scientific career, identifying its goals, locating its successes and failures.

Part IV explains why operant psychology needs to abandon (indeed, has already abandoned) radical behaviorism, then locates operant psychology within psychology as a whole. It argues that the major principles of operant psychology have nothing to fear from cognitive theory, and indeed, offer psychology's only source of answers to certain important questions. The final sections of the final chapter round off the discussion by asking whether operant theory has any significant ethical or political implications. The answer proffered is "Yes, but not the ones drawn by radical behaviorism."

Terry L. Smith
Takoma Park, MD
July 4, 1993

ACKNOWLEDGMENTS

The following material is based upon work supported by the National Science Foundation under Grant No. DIR-89121291, a 1989 Summer Stipend from the National Endowment for the Humanities (FT-32325), a 1989 and a 1991 Faculty Senate Summer Research Award from the University of the District of Columbia, and a 1989-1990 sabbatical grant from the University of the District of Columbia. The author gratefully acknowledges the support received from these institutions. The Government has certain rights in this material. Any opinions, findings, and conclusions or recommendations expressed in this material are those of the author and do not necessarily reflect the views of the National Science Foundation, the National Endowment for the Humanities, or the University of the District of Columbia.

I want to express my gratitude to Dudley Shapere, Charlie Catania, Ron Amundson, Jim Joyce, Al Mosley, Jon Ringen, and Fred Dretske for their assistance. Each of them influenced this project for the better. I owe a special debt to Michael Schulman, Jon Ringen, Charlie Catania, Fred Dretske, and a blind reviewer for commenting upon an earlier version of this manuscript. Those comments guided revisions of the NSF report (T. L. Smith, 1991) upon which this book is based.

I acknowledge the B. F. Skinner Foundation for permission to use two figures from Ferster & Skinner (1957), and the American Psychological Association for permission to use a figure from Skinner (1956). I thank Rob Jones for technical assistance with my computing problems, Manon Cleary for permission to use a copy of her oil painting for the cover illustration, Richard Colker for commenting upon and proofing the penultimate draft, and Jim Fetzer for steadily encouraging me across the finish line. Finally, I want to express my warm appreciation to my wife, Nora Blue, who has supported this project in numerous ways even as she launched a career and shared in the care of Russell and Trisha.

ACKNOWLEDGMENTS

INTRODUCTION

THE ANOMALOUS SURVIVAL OF OPERANT PSYCHOLOGY

> Skinner's program, unlike other behaviorist programs, has continued to develop unabatedly for the past forty years. (Lacey, 1979, p. 381)

The conventional wisdom of the 1970's was that during the previous decade psychology had undergone a revolution. The precise nature of this revolution was a matter of dispute, but the revolution itself was not. Some described it with concepts introduced by Thomas S. Kuhn (1970), suggesting there had been a "paradigm shift" from behaviorism to cognitive psychology (Leahey, 1980). Others argued that there had never been a behaviorist paradigm, but only a behaviorist methodology, and so there could not have been a paradigm shift, but only a decisive rejection of a failed methodology (Mackenzie, 1977). Still others suggested that Noam Chomsky's (1959) critical review of B. F. Skinner's (1957) *Verbal Behavior* had actually refuted behaviorism (Newmeyer, 1980), while others maintained that the story was more complicated than a straightforward case of refutation would allow (Lachman, Lachman, & Butterfield, 1979). No one however doubted that increasing numbers of psychologists ignored behaviorist strictures against reference to mental states and processes. Behaviorism, the dominant approach to psychology in America from roughly 1920 through 1950, seemed to be dead--or at least rapidly dying.

I

There was, however, one striking exception to this trend. In 1954, when William Verplanck (1954) compared B. F. Skinner's operant theory with other learning theories of the day, he found it had "uncovered a wide new range of phenomena, involving variables not at all considered by others" (p. 302). In 1959--ironically, the very year in which Chomsky allegedly refuted behaviorism--the number of articles published by operant psychologists actually began to show a dramatic increase (Gilgen, 1982, pp. 97-98). During the following decade, as cognitive psychology rose to its current prominence, operant psychology was also experiencing rapid growth. This growth can be documented by almost any measure one might choose--number of research articles, expansion of professional organizations, or influence on the field as judged by peers (Gilgen, 1982). Operant theory, however, is usually seen as embodying an extreme form of behaviorism. Thus, if most behaviorist psychologies were dying, one was nonetheless emerging in full bloom.

At first, this was not widely noticed; but as the dust from the cognitive revolution settled, and historians began to sift through the rubble of behaviorist psychology, they discovered an operant program that was fully intact and showing no signs of collapse.[1] Of necessity

[1]Catania (1973a, p. 434) seems to have been one of the few scholars to express serious doubts prior to the 1980's about there having been a cognitive revolution in psychology.

> The nineteenth century closed with the promise of an integrated science of psychology (Tichener, 1898). In the twentieth century, that promise has yet to be fulfilled. Students of psychology still are asked to choose theoretical sides. They see functional accounts of operant behavior pitted against ethological accounts of behavioral structure, analyses of reinforcement contingencies pitted against theories of cognitive processing, and descriptions of language as verbal behavior pitted against psycholinguistic formulations of language competence. Behaviorism continues to clash with phenomenology, and empiricism with nativism. Psychologists are

the history of psychology must be revised to accommodate this puzzling countertrend. Thomas Leahey, who was once a strong advocate of the paradigm shift interpretation of the cognitive revolution (Leahey, 1980), now says "it would be false to assert, as many have, that [the operant program] is dead" (Leahey, 1987, p. 461). He even denies that a cognitive revolution ever occurred, asserting that "there never has been a paradigm in psychology" (p. xiii), and reasoning that if there never has been a paradigm then there cannot have been a Kuhnian revolution. He furthermore says that if one had to identify a genuine paradigm in psychology, the operant tradition is "without doubt the closest" thing to it (p. 382).

None of this is to say that Leahey endorses radical behaviorism. He does not. His re-evaluation of the cognitive revolution, however, is indicative of a sea change in the interpretation of psychology's recent history. Now that we have a longer perspective, what once seemed like a winner-take-all contest that cognitive psychology won has begun to appear less decisive and more complex. Even those who continue to speak of a cognitive revolution are adding nuance to their descriptions. Baars (1986), for example, suggests that we should not view the era immediately preceding the cognitive revolution as having been totally behavioristic, nor the era immediately following it as

---

> not yet even agreed on whether theirs is a science of behavior or a science of mental life.
>
> The development of these controversies has been described in terms of paradigm clash (e.g., Katahn & Koplin, 1968; Neisser, 1972; Segal & Lachman, 1972), as if psychology were in the midst of the kind of scientific revolution described by Kuhn (1962). The student, whether his mentor be cognitive psychologist or behaviorist, is led to believe that one or the other paradigm will emerge victorious from the confrontation of incompatible intellectual positions. But this characterization may be misleading, because it is not clear that the controversies have grown out of incompatible treatments of common problems.

It would take the rest of us a decade or so to notice that there was less to the idea of a cognitive revolution than at first met the eye.

having been totally cognitivistic. Instead, we should acknowledge that both approaches have maintained a significant presence all along. The net effect of the revolution was a mere change in relative influence (pp. 407-408).

The Point of Departure. The facts that puzzle the historian also puzzle the philosopher. Cognitive science has become the dominant approach to psychology, behaviorism in most forms is a thing of the past, yet the most behavioristic research program of all survives, and even prospers. This paradox shall serve as the point of departure for our philosophical inquiry. How has it been possible for the operant program to survive? And what are the implications for the rest of psychology? These are the questions that shall occupy us throughout this essay. Our attempt to answer them requires an analysis of the Skinnerian program. What are its concepts and the mode of explanation in which they figure? What are its discoveries and their relation to the mentalistic theories of cognitive psychology? These are challenging questions. Although operant theory makes no reference to underlying states or processes, it purports to explain complex behavior showing signs of intelligence and/or purpose--the very behavior that cognitive psychology and common sense interpret as the result of the subject's underlying beliefs and desires. How are we to resolve this apparent contradiction? To arrive at an answer, we shall be forced to disentangle operant theory from the philosophy of behaviorism.

Behaviorism. In preparation for this undertaking, let us briefly discuss the concept of behaviorism. The term *behaviorism* can be defined in a number of ways, not all of which are relevant to our purposes. On the broadest definition, behaviorism is simply the assertion that psychology is about behavior (as opposed to consciousness or the psyche). It is sometimes argued that, on this broad definition, most (if not all) scientific psychologists are behaviorists, inasmuch as ultimately any psychologist is interested in explaining behavior (Leahey, 1992). This, however, is doubly misleading. First of all, even on this broadest of definitions, not all psychologists will turn out to be behaviorists. Cognitive scientists are interested in explaining (among other things) the behavior of organisms, but this does not mean that cognitive science is about the behavior of organisms. Behavior is simply one kind of evidence that can be brought to bear upon the nature of cognition. Even on this broad definition, then, it is not true that most psychologists are behaviorists. But the definition itself is misleading because it suggests that the central issue raised by behaviorism is the question of what psychology is about, whereas "the basic question is the usefulness of mentalistic concepts" (Skinner, 1969, p. 267).[2] Behaviorism started out as a reform movement within academic psychology. It asserted that if psychology would adopt the concepts and methods of the natural sciences, there would be steady progress towards positive knowledge (Watson, 1913). Behaviorism was not, however, simply a directive to use the scientific method. Most academic psychologists, whether behaviorists or not, wanted to follow scientific method. What separated behaviorists from other psychologists had less to do with a commitment to scientific method than with a conception of what the commitment to scientific method entails.

The central and defining feature of behaviorism has been a negative thesis to the effect that a truly scientific account of behavior may not refer to mental states--i.e., must avoid reference to states such as beliefs and desires (Rosenberg, 1988). This implies that psychology must divorce itself from common sense. For beliefs and desires are the fundamental tools by which common sense explains behavior. Why did David tiptoe past the door? Because he believed the cat was on the mat, and he did not want to disturb it. Why did George put the saw

---

[2]This discussion assumes that *mind* and *behavior* retain their conventional contrast. There were, however, some theorists (the so-called logical or analytical or philosophical behaviorists) who held that (appearances to the contrary notwithstanding) mental concepts themselves could be analyzed as dispositions to respond in physically describable ways to physically describable stimuli. They thought we could replace statements about mental states with statements about behavioral dispositions without loss of descriptive or explanatory adequacy. Had they been right, then psychology would have turned out to be about behavior after all, for consciousness, mind, cognition, etc., would have themselves been forms of behavior. It is clear, however, that such analyses cannot succeed (for reasons to be discussed in Part II, Chapter 5). So we can safely bury such abstruse possibilities in this footnote.

beside the shed? Because he desired that the tree be cut down, and he believed that making the saw available to the woodsman would further this outcome. Behaviorism would banish such explanatory devices from scientific accounts.

Equivalently, behaviorism is the injunction to ban states having propositional content from psychology. A belief, for instance, is always a belief that something is the case. David believes that the cat is on the mat. This belief has a content specified by the proposition that the cat is on the mat. George desires that the tree be cut down. This too has a content, specified by the proposition that someone cuts down the tree. One of the vexing questions of philosophy is how a state of a person can have such a propositional content. Such states are not found in the physical sciences. Ordinary explanations of human behavior, on the other hand, make constant reference to them. Behaviorism makes a clean break with common sense on exactly this issue. Different behaviorists have set forth different reasons for doing so: mental states are unobservable, cannot be measured independently of one another, are unnecessary for purposes of explaining or predicting behavior, do not really explain anything, and in fact do not even exist. The distinctions among the major versions of behaviorism have been based upon the various reasons offered for avoiding mental concepts. What they have had in common, however, are the determination to avoid reference to the mind, and the conviction that only by so doing can we arrive at a scientific understanding of behavior.

Contemporary philosophers are skeptical of behaviorism, no matter how defended. Joseph Margolis (1984), for example, includes a chapter on behaviorism in his *Philosophy of Psychology*, and arrives at the carefully worded conclusion that "at the very least . . . the principal forms of behaviorism appear inadequate to the tasks of an empirical psychology" (p. 47)--a statement which would easily win the agreement of most philosophers of psychology. Alexander Rosenberg (1988), in *Philosophy of Social Science*, draws a similar conclusion about the project of reorganizing the social sciences to avoid reference to inner states having propositional content. It simply "has not succeeded" (p. 79). Behaviorism may have developed "technologies for the control and prediction of behavior . . . in highly restricted settings," and forged "new tools" for experimental psychology (p. 79), but it has failed in the central task of providing a basis for a progressive science of human behavior. Margolis and Rosenberg do not offer idiosyncratic evaluations; they simply summarize the current state of reasoned opinion on these matters. The number of psychologists and philosophers who agree with Margolis and Rosenberg is large enough that theirs is the standard evaluation of behaviorism. This fact only hardens the paradox that we confront, for the theories and explanations of operant psychology seem to have won for themselves a place within psychology at odds with the standard evaluation of behaviorism. Evidently, we need a philosophical analysis of the operant program that will resolve this paradox.

II

Before beginning our philosophical investigation, we should acknowledge several objections that might be raised against the cogency of our project. We have taken as our point of departure an historical judgment that operant psychology has won a legitimate place for itself within psychology, and combined this with a philosophical judgment that operant theory explains intentional behavior without reference to mental states, to draw the modest conclusion there is something paradoxical about operant psychology that requires philosophical analysis. But what if one of our assumptions is false? What if the apparent success of operant psychology is illegitimate? Or what if operant theory does not really explain behavior or makes covert reference to mental states in doing so? Let us examine these possibilities and assess their relevance to our conclusion.

Are Operant Psychology's Successes Scientific? Operant psychology's survival and expansion during the cognitive revolution of the 1960's was in itself an impressive accomplishment. But this would not be the type of accomplishment that merits a careful philosophical analysis of the program if it can be traced to non-scientific factors. There are several reasons why one might suspect that non-scientific factors played a dominant role.

The program has isolated itself from the rest of psychology (Krantz, 1971, 1972). According to one observer it has "bypassed the mainstream of the American psychological establishment," and created parallel institutions of its own (Guttman, 1977). When journals rejected their articles, operant psychologists created their own journals. When the dominant professional organizations left them disenchanted, they formed their own. And more recently, when they began to despair of reforming the discipline of psychology, they even broached the possibility of defining a new discipline with degrees and departments of its own (Epstein, 1984; Fraley & Vargas, 1986).

Leahey (1987) claims operant psychology now finds itself in "a sort of publications ghetto" (p. 444). One critic has called it a "cult" (Wendt, 1949), and psychologist Irving Maltzman has questioned whether its growth is due to its scientific accomplishments.

> Because psychology isn't a highly developed science, much of what you call theories in psychology are very dependent upon a particular individual's charismatic qualities, his ability to attract people and to excite them. This is the reason for Skinner's success. He's a great PR man, and the reason there is such an interest in Skinnerian approaches to so many problems is not because there have been some great profound advances in knowledge, or some clearly established successes, but because the guy's a great publicist. (Maltzman, 1986, p. 103)

If Maltzman is right, then operant psychology's growth would pose only the historical problem of explaining how it occurred at the same time that other forms of behaviorist psychology were contracting. The program itself, however, would no longer require philosophical interpretation, any more than the continued existence of astrology (millions of people check their horoscopes before making important decisions, apparently including former President Ronald Reagan) induces philosophers to give careful scrutiny to zodiacal concepts.

But mere institutional success cannot account for the degree of interest psychologists have shown in operant ideas. Perhaps if this interest were purely critical, aimed at correcting what critics might see as unjustified influence, this would be consistent with Maltzman's claim. But the interest shown by other psychologists is sometimes sympathetic, taking the form of applying or borrowing ideas, rather than of criticizing them. Indeed, Gilgen (1982) found Skinner to be the most influential figure, and found Skinner's contributions to be the most important influence, in post-World War II American psychology, as judged by a random sampling of members of American Psychological Association, and also as judged by a random sampling of members of APA Division 26 (devoted to the history of psychology).

Furthermore, operant psychology's relative isolation can hardly have enhanced the program's influence. Although this isolation appears to be decreasing, it nonetheless continues to exist (Coleman & Mehlman, 1992). So if the work of operant psychologists has influenced other psychologists, this influence has probably occurred despite its isolation, not because of it. Furthermore, this isolation might in part be due to the sorts of misunderstandings that our analysis seeks to dispel. Thus, the isolation of the operant program does not undermine our conclusion, which is simply that there is something about this program that deserves sympathetic philosophical attention.

Is It a Mere Technology? A second line of rebuttal concedes that, yes, operant psychology has exerted influence beyond the confines of its own narrow circle, but suggests that this influence has been confined to the program's technological accomplishments. The design of the operant conditioning chamber, the study of responses the animal can easily perform and of stimuli the animal can easily discriminate, the use of conditioned reinforcers to maximize experimental control, the procedures for handling animals before and after experiments, these are all genuine (if modest) contributions to psychology (Mackenzie, 1977). Likewise, the application of operant concepts to the control of human behavior has proven to be effective, leading to the development of the technique of behavior modification, which Gilgen (1982) found to be second only to Skinner's own contributions in its influence upon members of the American Psychological Association. But these accomplishments may strike one as more technological than scientific (Gilgen, 1982, p. 297). They are useful for the control of behavior, but do they provide scientific understanding? Some psychologists have suggested the answer is no (Bolles, 1984). And once again, if operant psychology's success is not scientific, further investigation is unlikely to be philosophically rewarding.

We may begin our response by underlining the fact that operant psychology has indeed developed techniques of behavioral control. But this does not mean that control is the only, or even the principal, goal of the program. Enhanced ability to control events is a natural consequence of any science that refines our understanding of the causes of those events--so long, at least, as we can control the causes themselves. Therefore, operant psychology's success in controlling behavior is not necessarily an argument against its legitimacy as a science.

Indeed, quite the opposite. Donald Baer (1978), one of the founders of behavior modification, argues that technological applications of operant principles are scientifically significant because they test the generality of the principles upon which they are based. There is no guarantee that principles established under the controlled conditions of the laboratory will generalize to the uncontrolled settings of everyday life. And when they do, this redounds to the theoretical credit of those principles. Hence, the effectiveness of behavior modification is prima facie evidence that the same operant principles that explain the behavior of rats and pigeons in the laboratory chamber can also explain ordinary human behavior in everyday settings (cf., Schwartz & Lacey, 1982).[3]

Absent a philosophical analysis of the operant program, one does not know whether to infer that it is bad science hiding behind good technology, or that it spawns good technology because it is good science. Therefore, far from showing that a philosophical analysis is superfluous, the objection under consideration shows how much we need one.

---

[3]To say that operant principles explain behavior, whether of human beings or rats, is sure to evoke dissent from many philosophers and psychologists (including some operant psychologists, e.g., Catania, 1979). Permit me to issue a promissory note that entitles the reader to an account at some later point of the sense in which operant principles can be said to explain behavior. Parts of this note may be redeemed by turning to the discussion of the tautology problem in Chapter 4 and the discussion of operant theory's relationship to cognitive theory in Chapter 11.

Do Its Applications to Human Behavior Circumvent Mentalism? The third rebuttal acknowledges that yes, operant psychology has developed a genuine science of behavior, but then suggests that its application to human behavior implicitly relies upon mentalistic concepts. According to this line of criticism, when dealing with rats or pigeons in the confines of a standard experimental chamber, operant theory gives a genuinely nonmentalistic account. But when applied to human action in everyday settings, it unwittingly lapses into a mentalistic usage. This argument first appeared in Chomsky (1959), and was repeated in Chomsky (1971). By now it has become a standard feature of philosophical discussions of operant theory. Robert Audi (1976), for example, wonders whether an operant account of human behavior is not equivalent to an intentionalistic one that appeals to beliefs and desires (p. 178). And both Margolis (1984) and Rosenberg (1988) round off their discussions of operant psychology by suggesting that operant concepts may have surreptitiously taken on an intentionalistic (i.e., mentalistic) character. The implication is that insofar as operant psychologists have something valid to say about human behavior, it is because they have tacitly shifted the meanings of key terms so they no longer have their rigorous laboratory meanings, but are now roughly equivalent to concepts of folk psychology.

There is an ironic history to this critique. In the 1950's and 1960's, philosophers typically treated operant psychology as virtually an ostensive definition of the behavioristic approach to psychology. At the same time, however, Skinner grew increasingly insistent that operant psychology was not committed to certain doctrines traditionally associated with behaviorism. It did not, for example, assume that stimuli and responses could be defined in terms of their first-order physical properties, did not object to reference to subjective phenomena, was not based upon the philosophy of operationalism, did not reject reference to theoretical entities, etc. In place of the view that operant psychology adheres to an extreme form of behaviorism, Skinner (1969) suggested that it is an attempt to take a middle path between S-R psychology and mentalistic psychology (pp. 27-28). But middle paths are notoriously difficult to follow. And to some observers, it seemed that operant psychology had become so intent upon avoiding the excessively reductionistic approach of S-R psychology that it had strayed too far in the opposite direction and become implicitly mentalistic. If so, then its ability to offer a plausible interpretation of human behavior could be attributed to its use of intentionalistic concepts. It would then turn out to be an instance of the dominant trend towards mentalism in psychology. The principal difference would be that in this case the mentalism of the concepts is camouflaged by self-deception. Of all the criticisms of operant psychology, this is for operant psychologists the least expected and the most bewildering. Perhaps this is why they have never mounted a convincing response, even though it is now a longstanding challenge to the behavioral approach.

Actually, operant psychologists no longer rigidly dismiss mentalism as inappropriate to any form of psychology, but they still believe their own theory to be non-mentalistic. This theory however is difficult to understand--as Verplanck, Koch, Scriven, Lacey, Dennett, Ringen, L. D. Smith and others have testified. Thus, it is at least as plausible to think that philosophers have misunderstood operant theory as to think that operant psychologists have misinterpreted their own technical vocabulary. Here again, this difficulty of interpretation actually supports our contention, which is that we need a philosophical analysis of the operant program.

Summary. Although few philosophers take behaviorism seriously as a strategy for organizing the entire discipline of psychology, the anomalous success of operant psychology poses a challenge to current philosophy of psychology. The present work takes as its point of departure the fact that there appears to be an established theory of intentional behavior that has been making steady progress without recourse to mentalistic concepts. The major problem is to explain how such a theory is possible and to reconcile its claims with those of cognitive psychology.

PART ONE

UNDERSTANDING THE PROGRAM OF RESEARCH

> Skinner's conditioned responses seem to many readers just as *mere* as those of Pavlov or Hull, with the extraordinary result that he has been classed with Hull rather than with Tolman, with Guthrie rather than with Lewin, in his general position. Skinner's work has, in fact, very little in common with that of any of these men. (Verplanck, 1954, p. 307)

In the summer of 1950 seven psychologists attended a special seminar at Dartmouth College for the purpose of evaluating the major theories of learning. They chose to focus upon the theories of Clark Hull, Edward Tolman, B. F. Skinner, Kurt Lewin, and Edwin Guthrie. The results of their deliberations appeared in *Modern Learning Theory* (Estes, Koch, MacCorquodale, Meehl, Mueller, Schoenfeld, & Verplanck, 1954). Interestingly enough, none of the theorists met the seminar's standards for good science. In retrospect, this is not entirely surprising, since the seminar's standards were those of logical positivism. Of more importance for our purposes, however, is the fact that members of the seminar noticed that Skinner's approach to psychology, though superficially similar to that of the other theorists, was in fact profoundly different.

William Verplanck wrote the chapter on Skinner. He there notes that although Hull, Tolman, Guthrie and Lewin "share with us much of our view of theory," and therefore can fairly be criticized for failing to meet the seminar's standards, "it is not clear that Skinner is dealing with the same subject matter as these others" (Estes et al, 1954, p. 306).[1] Indeed his system seems to represent "a re-orientation towards" psychology with implications which are "in a sense, nihilistic" (p. 306).

> It proposes that all the conventional modes of thought in psychology, phenomenalistic, mentalistic, physiological, be rejected. It insists that psychologists begin their labors over again, that they develop their concepts from the ground up, and base them on the characteristics of the data themselves, and not on the language habits and intellectual biases of the theoretician. Earlier data may, where they meet the criteria of experimental control and orderliness of result, be salvaged, but earlier concepts may not. (p. 271)

Unfortunately, this aspect of Skinner's position is "too often overlooked" (p. 271). Skinner's frequent use of terms borrowed from other systems (but with quite different intended meanings) increases the likelihood of misinterpretation (p. 307). As a result, it requires a special effort just to understand what Skinner means by his scientific assertions.[2] This, then, shall be the topic of the following three chapters: understanding Skinner's concepts and the assertions he made with them.

---

[1]L. D. Smith (1986) gives reason to doubt whether Hull and Tolman shared so much of the seminar's view of theory as Verplanck seems to assume. But even taking these considerations into account, Skinner stands apart from the rest.

[2]Koch makes a similar point about the problem of interpreting Skinner. A member of the audience at the Rice University symposium of 1963 suggests that Koch's criticisms of behaviorism fail to join issue with Skinner. Koch concedes that Skinner is "the most subtle individual who has in some sense shared certain of the orienting attitudes of the behaviorists' point of view," and acknowledges that this makes it "rather difficult to cover Skinner exhaustively and, at the same time, talk about behaviorism in general." But then he adds that it is difficult to interpret Skinner under any circumstances, since "his position in connection with theory has always, it seems to me, been, shall we say, close to systematically ambiguous" (Wann, 1964, pp. 42-43).

CHAPTER ONE

DEFINING THE OPERANT

Skinner's initial goal for his research was a simple one: to arrive at a useful set of concepts to describe behavior. This may seem unnecessarily modest, but in his opinion psychology had attempted to move too quickly to an advanced stage of theorizing, skipping over a necessary exploratory stage during which it discovers quantitative relationships between independent and dependent variables (causes and effects).[1] Eventually, psychology would account for these relationships by means of hypotheses about inferred entities and processes. But to begin with, the problem was simply to find the relationships themselves. And to do this, it must find the correct descriptive categories.

I

In searching for such categories, Skinner did not start with a certain kind of behavior and ask what causes it, or with a certain factor and ask what it causes. Instead, he started with a few relatively well understood regularities, and then refined our understanding of them experimentally. In his early papers, he called such a regularity a reflex, by which he meant a causal relation between a stimulus and a response.

---

[1]Perhaps because of the influence of logical positivism on him, Skinner tried to avoid the use of causal language, but he seems eventually to have realized that this was not feasible, and by the 1950's was glossing independent variables as causes, dependent variables as effects (e.g, Skinner, 1953a, p. 35).

The Laws of the Reflex.[2] Skinner's analysis of the reflex emerged gradually, beginning with a paper (Skinner, 1931) drawn from the first half of his doctoral dissertation. In it he attempts to reformulate the physiologist's concept of the reflex so that it can be applied to the behavior of the intact organism.

An example of the type of reflex that the physiologist studies is the movement observed in the tail of a decapitated newt when its spinal chord is stimulated at a certain location. When physiologists discovered this phenomenon in the 18th century, they were puzzled. It violated their expectation that behavior is always the result of conscious, learned, voluntary mental activity. They immediately inferred various mechanisms to explain what they saw; but Skinner thinks their inferences were premature.

> In the history of the reflex one positive characteristic has always been given by the facts--the observed correlation of the activity of an effector (i.e., a response) with the observed forces affecting a receptor (i.e., a stimulus). The negative characteristics, on the other hand, which describe the reflex as involuntary, unlearned, unconscious, or restricted to special neural paths, have proceeded from unscientific presuppositions concerning the behavior of

---

[2]Skinner and Verplanck, following the practice of their time, referred to experimentally derived regularities as laws. Professor A. C. Catania objects (in a personal communication) to the use of the term *law* in relationship to behavioral regularities. He notes that the term implies a kind of universal applicability that one does not expect to find in biological or psychological generalizations.

Granting this point, I nonetheless do not think Skinner or Verplanck meant to imply universal applicability when they used this term. By calling a regularity a law, they simply meant to distinguish it from accidental generalizations that cannot support counterfactual conditionals. Indeed, all operant psychologists interpret the regularities they discover to be capable of supporting counterfactual conditionals. They would agree, for example, that if the rat had been exposed to a fixed-ratio schedule of reinforcement instead of a fixed-interval schedule, then its cumulative record would have a stair step pattern rather than a scallop pattern.

To avoid unnecessary controversy, I attempt to avoid use of *laws* where possible, but it would be awkward and misleading to do so in a discussion of the origins of operant concepts. By the time one gets into the 1970's and 1980's, however, the term is almost entirely absent from standard operant usage. I know of only two exceptions: *law of effect* and *matching law*.

> organisms. When Marshall Hall decapitated his famous newt, he pointed quite correctly to the reflex activity of the parts of the headless body, to the observed fact that movement followed, inevitably, the administration of specific stimuli. But his assumption that he had imprisoned in the head of the newt the source of another kind of movement was irrelevant and unsupported. The fact before him was a demonstrable necessity in the movement of the headless body; his failure to observe similar necessities in the movement of the intact organism was the accident of his time and of his capabilities. (Skinner, 1931, p. 331)

Skinner proposes that a reflex tentatively be defined as simply "an observed correlation of stimulus and response" (Skinner, 1931).

The other properties that historically have been associated with the reflex (involuntary, unlearned, unconscious causes restricted to special neural pathways) should be dropped from the definition. They constitute inconclusive speculation about what underlies (or does not underlie) the correlation between the two end terms. One can always go on to investigate this question (as the science of physiology does), but this does not affect the reality of the correlation itself. So far as the study of behavior is concerned, the basic object of study is the correlation itself. The study of the reflex divides into two parts. The first part studies laws of a given reflex at a fixed point in time. It addresses quantitative issues such as the following: How much time passes between the stimulus and the response (latency)? How strong does the stimulus have to be before it elicits a response (threshold)? How does the strength of the stimulus affect the strength of the response (magnitude)? And so on. Its laws are of the form

$$R = f(S)$$

where R is a response, S a stimulus, and f a mathematical function mapping some aspect of the stimulus onto some aspect of the response. Skinner calls these the static laws of the reflex. Their values at a given moment provide a measure of reflex strength.

The second part studies changes in these static properties of the reflex as a function of third variables such as number of responses (fatigue), drive (hunger), emotion (anger), or experience (conditioning). Its laws are of the form

$$R = f(S,A)$$

where A is any variable that may help to account for changes in the value of R relative to S. Skinner calls these the dynamic laws of the reflex.

Dynamic laws are especially important for study of the intact organism. Surgical isolation of the two halves of a reflex is now (by definition) impossible. Furthermore, the number of reflexes of the intact organism is virtually infinite, so the classification of individual reflexes ("botanizing") will be of virtually no value (Skinner, 1938). The science of behavior must therefore focus upon laws of the second type.

Defining the Operant. Although Skinner's early concept of the reflex seems to fit the pattern of physiology, appearances are deceiving. E. G. Boring, who was one of the readers on Skinner's dissertation committee, made an astute observation about Skinner's use of the term *reflex*.

> You are making an argument for keeping the word *reflex* and giving it a new, broader, and relatively strange meaning. No one would guess this to be your goal as you start in, and you yourself may not think of it in that way.... You have given a very broad, strange, almost bizarre meaning to the word *reflex*. You have taken it away from the constrained anatomical reflex-arc meaning and you have equated it to the concept of psychological fact-as-relational-correlation which already has terms for itself. What is the use? To wrench the word from its well-entrenched meaning, you need more than a paper; you need propaganda and a school. And if you succeeded you would have merely an equivalent for Gestalt with a special epistemology back of it. (Skinner, 1979, pp. 72-73)

Skinner must have to some extent agreed, because he would eventually restrict the meaning of reflex to coincide more nearly with Boring's. But this would occur only after he had introduced a special term to refer to a very unreflexive kind of behavior.

Skinner's definition of the operant begins with an informal (i.e., non-experimental) description of a certain broad behavioral regularity. Skinner worked his way toward an understanding of this regularity over the course of several years. In his autobiography, he writes that by the end of 1932 he was beginning to realize that the role of stimuli in relation to responding was itself an interesting question.

> The orthodox question was whether an organism could see something--say, a very faint light or the difference between two colors or patterns. I was not interested in capacity but in the role played by the stimulus. It was becoming clear that the light did not *elicit* the response in the sense in which a tap on the patellar tendon elicits a kick of the leg, nor was the lever simply a collection of sights, smells, and touches having that effect. Of course, the lever stimulated the rat before a response was made and reinforced, but its effect was upon the probability that pressing would occur. As in my treatment of drive, I was breaking away from the traditional view of a stimulus as a goad. (The two concepts were combined by psychologists who included drive in a "total stimulus situation.") The temporal order of stimulus and response suggested causal action, but it was not the action of a force. (Skinner, 1979, p. 105)

Instead of asking what are the discriminative or behavioral capacities of the organism (the basic question addressed by the botanizing of reflexes), he asked what roles are played by such capacities (whatever these capacities may be for a given organism at a given stage in its development). And he began to see that there are two quite different roles these capacities can play.

Skinner's first written attempt to make the distinction between two types of reflex came in a 1935 article, "Two Types of Conditioned Reflex and a Pseudo-Type." He observed that one type of conditioning takes place when a reinforcing stimulus (e.g., food) that is capable of eliciting a response (e.g., salivating) occurs in conjunction with the onset of some other stimulus (e.g., a tone). The result is that the latter stimulus (the tone) comes eventually to cause the occurrence of the response (salivating). This is the Pavlovian or classical form of conditioning. Schematically, he represented this type of conditioning as follows.

Tone -- Food → Salivation
S S R

leads to

Tone → Salivation
S R

(a dash represents co-occurrence,
an arrow represents causation, and
S & R represent stimulus and response)

The other type of conditioning takes place when a reinforcing stimulus (e.g., food) occurs in the presence of some other stimulus (e.g., the lever itself) in conjunction with the occurrence of some response (e.g., a lever press). The result is that the latter stimulus (the lever) comes to cause the response (the lever press). This is the instrumental form of conditioning.

Lever -- Lever Press → Food
S R S

leads to

Lever → Lever Press
S R

The basic distinction between the two forms of conditioning is that in the classical case the reinforcing stimulus (food) is paired with a stimulus, whereas in the instrumental case it is paired with a response.

Skinner interprets the two kinds of conditioning as playing different roles in "the economy of the organism" (Skinner, 1935b, p. 375). The role of classical conditioning is to substitute one stimulus for another. It thereby "*prepares* the organism by obtaining the elicitation of a response before the original stimulus has begun to act." As a result of classical conditioning the organism begins to salivate even before food is delivered, thereby preparing itself to ingest the food more efficiently.

Instrumental conditioning, on the other hand, "selects from a large repertory of unconditioned reflexes those of which the repetition is important with respect to certain elementary functions and discards those of which it is unimportant." A response which is first emitted as part of an "investigatory reflex" is effective in producing food, and as a result of instrumental conditioning the response gains in strength, thereby making the organism more effective in producing food in the future.

A few years later, Skinner would write that there are not only two kinds of conditioning, but two kinds of behavior prior to conditioning. This bifurcation is roughly equivalent to the common sense distinction between voluntary and involuntary behavior. Voluntary behavior consists of "responses uncorrelated with observable stimuli" (Skinner, 1937, p. 378). This type of unconditioned behavior contrasts with the type of unconditioned behavior Pavlov studied. The latter "is made to specific stimulation, where the correlation between response and stimulus is a reflex in the traditional sense." Skinner coins the term *respondent* to refer to this type of behavior, thereby suggesting that it is always a response to some form of stimulation. Examples are the dog's salivating response to the smell of food, the rabbit's eye blink response to a puff of air, or the cat's flexion of the leg in response to a noxious stimulus.

The other type of response--Skinner's continued use of the word *response* for this type of behavior is an example of the sort of usage that has created so much misunderstanding--"occurs spontaneously in the absence of any stimulation with which it may be specifically correlated" (Skinner, 1937). In other words, this is a response without a stimulus. He refers to this type of behavior as *operant*, to suggest that its identifying property is the way it operates upon the environment. Operant behavior first appears in the form of spontaneous undifferentiated behavior not elicited by the environment.

Respondent behavior, on the other hand, comes in the form of unitary responses elicited by an identifiable part of the environment. Since conditioning cannot add new elementary respondents to the organism's repertoire, classical (respondent) conditioning can create new responses only by somehow bundling elementary responses together. By contrast, Skinner describes operant behavior as occurring spontaneously in undifferentiated form, and notes that this eliminates the need to assume that all operant responses exist "as identifiable units in unconditioned behavior"--or the need to assume (alternatively) that novel operant responses are created by combining identifiable unconditioned operants (p. 380). Instead, "elaborate and peculiar forms of response may be generated from undifferentiated operant behavior through successive approximation to a final form" (p. 381).

Skinner illustrates this process, which he later would call shaping, by describing a method for teaching a rat the response of pressing the lever.

> A rat may be found (very infrequently) not to press the lever spontaneously during a prolonged period of observation. The response in its final form may be obtained by basing the reinforcement upon the following steps in succession: approach to the site of the lever, lifting the nose into the air toward the lever, lifting the fore-part of the body into the air, touching the lever with the feet, and pressing the lever downward. When one step has been conditioned, the reinforcement is withdrawn and made contingent upon the next. With a similar method any value of a single property of the response may be obtained. (pp. 381-382)

Skinner implies that this process accounts for the acquisition of virtually all unitary operant responses (the sole exception being the assumption of operant control over a response that first appears as a respondent).

II

When asked why his investigation of the operant has been so successful, Skinner rarely mentioned his conception of the causes of behavior, and emphasized his conception of the effect. According to Skinner, "Progress in a scientific field usually waits upon the discovery of a satisfactory dependent variable" (Skinner, 1950, p. 46). In the case of chemistry, for example, knowledge began to accumulate only "when people were willing to disregard the very obvious and easily manipulated properties of compounds and substances and pay attention to the less obvious property of combining weight" (Wann, 1964, p.

100). Elsewhere he noted that "the science of mechanics moved forward rapidly when it was discovered that distances and times were more important for certain purposes than size, shape, color, hardness, and weight" (Skinner, 1953, p. 41). The implication is that a similar discovery would lead to progress in behavioral psychology.

The Dependent Variable of Operant Psychology. Skinner was the first psychologist to pay close attention to the rate of behavior.[3] One could make a case that his decision to focus upon rate was more important than his definition of the operant. The operant itself was not a totally new subject matter. Decades earlier Thorndike had studied something quite similar, if not identical, under the heading of instrumental behavior. But Thorndike was interested in how long it took an organism to learn the behavior. Skinner, on the other hand, was interested in how often the response would be performed once it had been learned.

This is not an obvious property of behavior. We do not easily discriminate changes in the rate of behavior unless we make use of special equipment. Thus, rigorous study of this property became possible only upon Skinner's invention of the cumulative recorder, which graphically summarizes the changing rate of responding over an extended period of time--often several hours. The effect of such equipment is "similar to the resolving power of the microscope" in the sense that "a new subject matter is open to direct inspection" (Skinner, 1963b, p. 111).

There was no guarantee, however, that such an inspection would find anything of interest. The mere fact that there is a quantitative dimension to rate does not mean that one can isolate the factors that control this quantity. Indeed, Skinner himself wasted a great deal of time studying not only rate, but also the number of responses an animal emits. For some time Skinner thought that conditioning builds up something he called the operant reserve (Skinner, 1938, p. 229). The

---

[3]As he had written in *Behavior of Organisms*, "the main datum to be measured . . . is the length of time elapsing between a response and the response immediately preceding it or, in other words, the rate of responding" (Skinner, 1938, p. 58).

operant reserve has as one of its dimensions the number of responses eventually to be emitted. The idea was that as you reinforced a response, you built up a reserve of responses which would then be emitted when reinforcement was curtailed. Certain aspects of how reinforcement was delivered were supposed to influence the size of the reserve. Other things--most notoriously, punishment--were not supposed to influence its size. If these alleged discoveries had held up under rigorous experimental scrutiny, they would have revolutionized our understanding of behavior. They would have meant, for example, that once an operant response has been reinforced, punishment of it has no effect on the number of times the response will be repeated. Punishment might temporarily suppress the response, but eventually the reserve would empty itself. So one of the principal justifications for punishment (viz., to prevent the commission of similar actions by the punished individual) would have been undermined. In fact, however, Skinner's early research on punishment was later overturned by some of his own students, and in general, work on quantity of responses never went anywhere. This is not to say one cannot show that certain factors influence the number of responses emitted under extinction--one can. But the results have not built on one another in the way results about rate have.[4]

Dynamic Laws of Behavior. Why is having this type of dependent variable so important? Because Skinner wants to measure something that changes value continuously as the result of changes in the value of something else. In terms of Mill's traditional methods of scientific discovery, Skinner wants to apply the method of concomitant variation.

---

[4]As Catania (1979) has noted, rate is not the only aspect of the operant response that contingencies of reinforcement are capable of controlling. Thus, if reinforcement is contingent upon achieving a certain force, duration, accuracy, etc., of responding, this aspect of the response can become the dependent variable. The key to the dependent variable is that it be free to vary more or less continuously as the experiment proceeds. This is what is meant by the free operant. The animal is free to operate upon the manipulandum (the lever, key, strap, etc.) repeatedly without physical restraint, artificially imposed delay, or any other form of interference. This insures that one achieves a numerical value that changes continuously throughout the experiment.

The other Millian methods of scientific discovery tell us what causes what, but concomitant variation tells us the quantitative relation between cause and effect. Skinner is not seeking to discover if making delivery of food contingent upon performance of a certain response will cause a hungry animal to perform that response more often. Obviously, unless there is an inherent incompatibility between the required response and the process of preparing to eat (as would be the case if the required response is to hold still, which a hungry animal about to be fed will find it difficult to do), a hungry animal will do what it has to do to get food. What Skinner seeks to discover are the quantitative aspects of such a causal relation: How much is the animal willing to do, and how rapidly will it do it?

Skinner succeeded in formulating the study of action in a quantitative way. This was to some extent a technological achievement. He invented the experimental chamber, making it possible to keep track of the continuous changes in some physical property (e.g., rate) of an operant response while carefully controlling the quantitative value of the variables that may have an effect on this property. It also, however, was a scientific achievement. Others tried to find quantitative principles of voluntary behavior, but they did not set up the problem in a way that led to a steadily developing program. Tolman, for example, focused on behavior at a choice point in a maze, but by the time the animal gets back to the same choice point again, so much has happened that it is difficult to establish causal relationships without a great deal of theoretical and statistical manipulation. Hull was interested in quantitative principles too, but he found them in the processes underlying behavior. These underlying processes, however, cannot be measured independently of psychological theory. This is not necessarily an insurmountable problem, but it does imply there is much less opportunity to refine such principles experimentally. One has to invent them through a creative hypothesis, then laboriously check to see if experiment bears them out. Since the experimental result is always the outcome of the interaction of several hypothetical processes, an experiment cannot directly confirm any given principle. Skinner did not reject such an approach categorically. He held only that there is a certain cost involved in such an approach. Where no alternative exists such costs are, he acknowledged, unavoidable (Skinner, 1950).

But he thought it is possible to study voluntary behavior one cause and one effect at a time, and measure the quantitative relationship between the two. He assumed that the advantages of such an approach during the early stages of investigation are obvious.

Inferring Hypothetical Entities. There is a widely held misconception that Skinner objected in principle to hypothetical entities, but he did not (Meehl, 1986). What he objected to was premature recourse to hypothesis. He thought sustained scientific progress requires quantitative principles arrived at experimentally through manipulation of physical variables that can be measured independently of one another. Drawing inferences about hypothetical entities before building an inventory of such principles would be more likely to slow the rate of progress, rather than to increase it. For it would invest valuable scientific resources in a project that would reap minimal results, whereas those same resources could have reaped larger returns if invested elsewhere.

CHAPTER TWO

NOT A FORM OF S-R PSYCHOLOGY

Skinner has long emphasized that his experimental program is not a form of stimulus-response psychology. There is a straightforward and widely recognized sense in which this is so: an operant response does not require a stimulus. In the simplest case, the organism already has a certain operant response (say, lever pressing) in its repertoire. It emits this response at a certain rate spontaneously, without benefit of stimulation. If we now arrange for delivery of a reinforcing stimulus (say, food) to be contingent upon this response, then conditioning will occur and the rate of the response will increase. But there is no sense in which the delivery of reinforcement after a given response elicits, in the manner of a reflex, the next such response. And in this sense, the response occurs without a stimulus. The effect of the environment is not to make the response occur (there can be an operant response without a stimulus), but simply to raise the rate of such occurrences. Respondent behavior, on the other hand, requires an environmental event if it is to occur--i.e., there is no response unless there is a stimulus. Thus, if we identify S-R psychology with the study of respondent behavior (classical, Pavlovian conditioning), then operant psychology will not be a form of S-R psychology.

This however is a philosophically insignificant way to draw a distinction between the operant program and S-R psychology. The goal of this chapter is to explicate a distinction that makes a difference.

I

Instead of making contact with a subject matter by focusing upon a certain type of cause or a certain type of effect, Skinner starts with a few causal relationships. The first of these is the pattern of respondent conditioning, in which a neutral stimulus that is paired with a reinforcing stimulus comes to elicit the same response as the reinforcing stimulus. The second is the pattern of operant conditioning, in which a reinforcing stimulus that is paired with a certain response comes to control the frequency of that response. Operant conditioning is distinctive in that the environment controls a response without this control being exerted by a stimulus immediately preceding the response. The animal presses the lever, a reinforcing stimulus is delivered, and the rate of lever pressing increases. The presses now occur at a higher rate than they did prior to delivery of the reinforcing stimulus. Some presses however would have occurred even in the absence of the reinforcing stimulus. On this description, operant conditioning constitutes a break with the stimulus-response principle that every response requires an eliciting stimulus. Operant responses do not seem to have such a stimulus.

Blurring the Distinction. The distinction between operant and respondent behavior is not, however, so clear-cut as this portrayal makes it seem. The environment can take on a function with respect to operant behavior that bears a strong resemblance to the role it plays with respondent behavior. For example, let a light in the experimental chamber come on when, and only when, lever pressing will result in the delivery of food. The rate of lever pressing is soon higher in the presence of the light than in its absence. Eventually, the light comes to control lever pressing to the point that when the light comes on, the animal begins to press the lever at a high rate, and when the light goes off, the animal stops pressing altogether. Operant psychologists say that the light has come to function as a *discriminative stimulus* and lever pressing has come under *discriminative control*.

The discriminative stimulus shares an important feature with the conditioned stimulus of respondent conditioning: A feature of the environment that once had no control over behavior was correlated with reinforcement, and the onset of this feature now causes responding. If the defining feature of S-R psychology is that every response must have a stimulus, then the difference between operant psychology and S-R psychology may be slim indeed. Although it may be true that unconditioned operant behavior can occur without a stimulus to goad it, once the behavior comes under discriminative control, there is a stimulus that precedes responding. Furthermore, in the absence of that stimulus, responding ceases almost altogether. A discriminated operant is therefore not so different from a respondent.

Indeed, even though Skinner thought unconditioned operants occur spontaneously without environmental stimulation, we cannot rule out the possibility of subtle controlling stimuli. Skinner seems to have been aware of such a possibility, for he was careful to say unconditioned operant behavior has no apparent stimulus, which by implication means he thought it may in fact have an *un*apparent one.

Skinner's early papers nonetheless drew a robust distinction between operant and respondent behavior. Operants can supposedly occur spontaneously, whereas respondents require stimuli. He furthermore tentatively suggested that operant conditioning applies only to skeletal responses whereas respondent conditioning applies only to autonomic responses (Skinner, 1938). Thus, there would be a different set of muscles for the two systems of behavior. The latter suggestion has proven untenable. Autonomic responses such as heartbeat, blood pressure, or unconscious tics have been shown to be subject to operant conditioning.

Perhaps more surprisingly, it also has been discovered that certain complex skeletal responses are apparently subject to respondent conditioning. The discovery of autoshaping, in particular, has considerably blurred the distinction between operant and respondent behavior. In the autoshaping procedure, a hungry pigeon is given access to food just after a light flashes behind the key in the pigeon's chamber. If repeatedly exposed to this procedure, the pigeon will eventually come to peck the lighted key, even though pecking does not cause the delivery of food. Furthermore, if the bird is in one part of the cage when the light comes on, it will move into position in front of the key and then start pecking. This is not the sort of behavior that used to be thought of as respondent, but the pattern of learning fits the pattern of respondent conditioning. On the Pavlovian paradigm, a bell rings and a while later the reinforcing stimulus is delivered, thereby causing the animal to salivate. After a number of repetitions of this pattern, the dog begins to salivate when the bell rings. Salivating has been conditioned to occur in response to the bell. Autoshaping follows the same pattern, with the difference that the response (pecking) is directed at the conditioned stimulus (lighted key), and the animal will perform other responses (moving across the cage) to make this response possible. These differences, however, blur some of the traditional distinctions between operant and respondent behavior.

Ultimately, the only sound basis for the distinction may be the procedural one with which Skinner began: with operant conditioning, the reinforcing stimulus is caused by the response it conditions, whereas with respondent conditioning, the direction of causation is reversed. For the sake of argument, suppose this is so. Suppose, furthermore, that operant responses always do require a stimulus, so that operant conditioning always involves transferral of discriminative control from one aspect of the environment to another. Hull seems to have been committed to something of the sort, in the sense that he found it economical to analyze instrumental conditioning as a special (more complex) case of classical conditioning. So there are theoretical motives, as well as empirical ones, for blurring the operant/respondent distinction. Would eliminating this distinction also eliminate the distinction between operant and S-R psychology? Does the distinction between Skinner and the S-R psychologists depend upon the principle that every response requires a stimulus? There are a couple of things wrong with such an analysis. First of all, it ignores the fact that the Skinnerian program has always included both operant and respondent conditioning within its purview. Did Skinner think it became a form of S-R psychology whenever it dealt with respondent conditioning, and then shifted back to a different approach when returning to operant conditioning? Skinner, I think, would have found this to be an odd characterization of his program. Second, and more importantly, such an analysis (mis)represents what is a conceptual difference as if it were an empirical one--as if certain types of empirical discovery would significantly narrow the gap between S-R psychology and Skinnerian psychology.

The problem, it seems, is that we are working with a relatively superficial analysis of what S-R psychology is. On a deeper analysis of its nature, there is a way of conceiving both operant and respondent conditioning so they fit the pattern of S-R psychology, and another way of conceiving them that does not. What distinguishes Skinner's approach to psychology is not so much its focus upon operant behavior, but its way of defining any form of behavior--operant, respondent, or otherwise.

A Deeper Conception of S-R Psychology. Charles Taylor (1964) has suggested that the basic question raised by behavioral psychology is whether behavior can be described on the basis of causal principles of a certain type. Taylor quotes Hull as saying that "an ideally adequate theory even of so-called purposive behavior ought . . . to begin with colorless movement and mere receptor impulses as such, and from there build up step by step both adaptive and maladaptive behavior" (p. 114). Taylor takes this to be definitive of a certain approach to psychology.

> This insistence on the kind of connection between "receptor impulses" and "colorless movement" is the essential principle of S-R theory and is what has earned it this name. The question, then, is whether causal principles linking events of this kind can be discovered which will account for behavior. (p. 115)

On this definition, S-R psychology says one should study what happens at the surface of the organism, where various forms of energy are transformed into afferent nerve impulses, and use this information to give a causal account of the way the animal subsequently moves its body in space.

S-R psychology takes exactly the opposite approach to making contact with a subject matter from Skinner's. It begins with a definition of the independent and dependent variables and assumes that regularities between them can be discovered. Skinner begins with known regularities and assumes that the end terms of the regularities can be defined at the same time as we refine our understanding of the regularities themselves. *A priori* there is little to recommend one approach over the other. In some respects the S-R approach may seem philosophically more profound, because it resolves certain metaphysical problems in advance. It requires that both independent and dependent variables be interdefinable with physiological constructs: stimuli would consist of the physical energy activating the transducers that produce afferent nerve impulses; responses would consist of the physical movements of the organism within the frame of reference of its immediate surroundings, as produced by a sequence of contractions of various muscles. The stimuli and responses of behavioral psychology would thereby be guaranteed to fit neatly into the ontology of the physical sciences. This indeed seems to be the philosophical motivation for this otherwise puzzling constraint.

S-R psychology takes a certain approach to causality. This approach is not limited to respondent conditioning. The essential feature is not that every response have a stimulus, but that every variable be defined in a way that relates to the physiology of the organism. The variables controlling operant responding need not be stimuli immediately preceding the response, but the causal principles of operant psychology would still fit Taylor's pattern if the variables (whatever they may be) are definable in this way.

It is easy to see why a behaviorist might adopt this approach. If psychology is to follow the example of the natural sciences, the variables entering into causal regularities must be defined as physical properties, physical forces, physical events, etc. The S-R approach reasons that there is already a science (physiology) that is working on the problem of how physical inputs give rise to physical outputs, so an obvious strategy for insuring that psychology will stay within the constraints of the natural sciences would be to require it to coordinate its concepts with physiology. Starting from a shared set of concepts, psychology would then tell the outside story while physiology tells the inside story. Psychology would search for causal principles that relate receptor impulses to colorless movement, and physiology would explain how these causal relations are mediated by the central nervous system. Physiology would then be able to pick up exactly where psychology leaves off, and the two would be complementary parts of some larger, unified scientific project.

This seems to have been the way Hull, behavioral psychology's most influential figure during the heyday of behaviorism in the 1930's and 1940's, conceived of psychology. But there are significant differences between psychology in the Hullian mode and psychology in the Skinnerian mode.

The Generic Concept of Stimulus and Response. Skinner's decisive break with S-R psychology came early in his career, in "The Generic Nature of the Concepts of Stimulus and Response" (Skinner, 1935a), written before he had drawn the operant/respondent distinction. In that article, Skinner proposes that we use dynamic laws to provide an experimental criterion of behaviorally real stimuli and responses.[1] Such a criterion can result in a stimulus not definable in terms of receptor impulses and a response not definable in terms of colorless movements. Instead, stimuli and responses are defined in whatever way will produce smooth curved dynamic regularities. In practice the end terms will often be described in the vocabulary of a "naive realism" that "rejects even the logical possibility of a reductionism" of the sort S-R psychology adopts as an *a priori* constraint (Verplanck, 1954, p. 308).

The empirically real units of behavior must be individuals that persist through time, that maintain their identity even while their properties change. The common sense world of things is populated by tables and chairs, plants and animals. These are fairly easy to identify and study. But what is the world of behavior populated by? What plays the role of tables and chairs or plants and animals?

In 1935, Skinner calls them *reflexes*. (He would later simply refer to them as *units of behavior*.) At this point in the development of his thought, he assumed that every response has a stimulus. A reflex is by definition a correlation of a stimulus and a response. The question is: what stimulus and what response? Even in the case of a simple reflex such as leg flexion, where a noxious stimulus applied to the leg causes retraction of the leg, it is not immediately clear to what extent we can arrive at well defined units which "retain their identity from experiment

---

[1]See note 2 of the preceding chapter on the use of the term *law*. As behavior analysis evolves, usage of this term changes.

to experiment" (Skinner, 1935a, p. 347). One's goal should be to define the stimulus and response in conformity with "the natural lines of fracture along which behavior and environment actually break" (p. 347). Skinner's problem is how to find these natural lines of fracture.

The stimulus that upon a given occasion causes a response, and the response which that stimulus causes, are two particular events. The unit itself cannot be this pair of events, however, for they will never recur. The unit must be something these events belong to. As a first approximation, Skinner considers the thesis that a reflex is a class of physically defined stimuli paired with a class of physically defined responses. Note that this is not the same as the S-R thesis that stimuli be receptor impulses and responses be colorless movements, because there are many possible levels of physical analysis besides those provided at the physiological level. The physiological concept of stimulus and response is a natural limit to the physicalistic approach. The maximally restrictive concept of the stimulus is just the physical energy that activates the sensory apparatus on a given occasion (or alternatively, the specification of the afferent impulse that the transducers are stimulated to produce); and the maximally restrictive concept of the response is just the topography of the organism's movements in space (or alternatively, the sequence of muscle contractions that produce these movements). On this conception, a reflex could be described in the language of physiology.

Skinner recognizes that unless there is some principled way to define the unit of behavior more inclusively (i.e., generically), we will end up defining it physiologically. For it would be difficult to stop short of the physiological level without being arbitrary. Thus, the quest for a generic conception of the reflex is equivalent to the quest for an alternative to the physiological elementalism of S-R psychology. At stake is not merely the issue of how to describe behavior, but also the issue of what behavioral psychology is responsible for explaining. For an immediate implication of the physiological definition of the reflex is that virtually any observable reflex would be a complex behavioral entity consisting of many S-R units bundled together by one or more relations or processes. In the case of the unconditioned reflex, these relations or processes are presumably innate. In the case of a conditioned reflex, some of the S-R pairs in the bundle would presumably get there by the process(es) of learning. The S-R approach thus assigns behavioral psychology a specific problem--namely, how to synthesize the observed units of behavior out of elementary S-R units.

At the time Skinner was writing, there were very limited possibilities for solving this problem. There was the process of chaining, which would bundle elementary responses together to form new responses. There was the process of association, which would link previously neutral stimuli with established responses. And there was induction, which would operate upon a stimulus that got tied to a response by association, and include in the bundle a number of similar stimuli that had never co-occurred with the response (and thus could never have been included in the reflex by means of association). S-R psychologists thus almost inevitably assumed that these processes accounted for the acquisition of new units of behavior (i.e., for virtually everything one might describe as learning). Thus, what Skinner characterizes rather blandly as the alternative to the generic view was in fact the dominant approach to academic psychology in the U.S. from roughly the 1920's through the late 1950's.

II

Skinner has said that at the most general level, his goal is simply to describe what organisms are doing. This may seem like a trivial task. Obviously animals run around, press levers, eat food, hear buzzers, see lights, and so on. But the problem of description becomes less trivial when we wish to use our descriptions to formulate causal regularities showing the effect of various circumstances on behavior over time. The physical stimulus and the physical response--i.e., the stimulus and response as S-R psychology defines them--do not necessarily give us the appropriate categories for describing such causal regularities. Neither does common sense.

A Unit Somewhere Between Physiology and Common Sense. Skinner notes that although one can usually specify in a loose way the defining properties of a response, one finds upon closer examination that certain non-defining properties of it make a difference in the dynamic laws.

If for example one sizably increases the force required to depress the lever, or raises the height of the lever significantly, or replaces the lever with a much smaller or a much larger one, then responses performed prior to the change would not necessarily be quantitatively interchangeable with responses performed after it. Thus changes in the non-defining properties of the reflex can disrupt the smooth curves of secondary processes, and so we cannot say these properties are completely irrelevant.

As a result, we must impose various restrictions upon our preparation. We must specify that the force required to depress the lever, the height of the lever, the size of the lever, and so on, all fall within certain limits. As we narrow these limits, the orderliness of the secondary processes increases. If we start out permitting the force required to depress the lever to double occasionally, this will disrupt the smoothness of our extinction curves. But as we narrow the parameters within which the required force may change, the dynamic laws become increasingly regular. Long before we have restricted our preparation to the maximum degree (at which point we would have completely specified the description of the stimulus and the response in the manner of S-R psychology), we reach a point at which the orderliness of the secondary processes quits increasing with further restrictions. This point "may appear at such a relatively unrestricted level--and, as one might say, so suddenly--that extrapolation to complete consistency appears to fall far short of complete restriction" (Skinner, 1935a, p. 359). In theory further improvement might be possible, but in practice the consistency of the secondary laws under a variety of circumstances is "so remarkable that it promises very little improvement from further restriction" (p. 359).

In fact, once we reach this point, the addition of further restrictions (e.g., requiring the rat to depress the lever with exactly a force of two grams) would have the effect of disrupting the secondary processes. Attempts to make things better now make them worse. We have encountered a behavioral limit to the usefulness of restrictive definitions. The same can be found on the stimulus side of the reflex equation. Keeping the lever from getting too hot or too cold may result in more regular dynamic laws.[2] But suppose we define the stimulus as a lever that is exactly 30 degrees Centigrade, and decide to exclude responses to any other stimulus from the reflex. This has a negative effect on the dynamic regularities. We will get smoother dynamic laws if we let such aspects of the stimulus vary somewhat.

There is a metaphorical sense in which the stimulus and response come into "focus" at a certain level.[3] Broadening or narrowing either stimulus or response class would decrease the orderliness of the secondary processes. Once having reached this level of maximal orderliness, "the problem of definition has now been practically solved" (Skinner, 1935a, p. 359). Skinner uses (at this point in the development of his thought) the term *reflex* to refer to such an experimentally defined unit.

On this usage, a term such as *lever press* will not refer unequivocally to a single reflex. For there are many different ways of restricting the preparation to attain behavioral focus. This however "is a necessary consequence of the complexity of the material, which cannot be changed by theoretical considerations" (p. 360). Thus, when writing about the lever press, it will be necessary to include a supplementary list of specifications such as the size of the lever, the pressure required to depress it, the height at which it is placed, and so on. This does not mean that all these restrictions are part of the definition of the lever press; just that the definition of lever press does not determine the experimentally real unit. "A rigorous definition without regard to non-defining properties is, in fact, probably impossible because, as we have seen, the defining property can be made to fail by taking extreme values of other properties" (p. 355). Such experimentally established classes will be narrower than the generic terms of ordinary language but broader than the terms of S-R psychology. Common sense would say the rat is pressing a lever, but Skinner would add that it is pressing a lever with a size and shape falling within certain limits, situated at a certain location, etc. Unless we add these qualifications, we are not

---

[2]Skinner at this point still analyzes a lever press as a response to a stimulus.

[3]Egon Brunswik (1952) would attempt to turn this metaphor into a philosophy of behavioral psychology (of which more below).

describing behavior in terms that correspond to behavioral causes and effects.

Of course, inasmuch as the type of generic stimulus Skinner is talking about cannot be specified in terms of the energy impinging upon the transducers of the organism, there is presumably a need to explain how the organism responds differentially to this stimulus. Something is happening inside the organism which makes this possible. Skinner does not claim to be able to say what this something is. But he does take an interest in specifying what the organism is responding to. His attitude towards the response end of the equation is similar. In the case of the lever press, for example, the rat is able to coordinate its movements so that they accomplish a certain result. If the lever is altered in certain ways, the rat's movements will adjust to attain the same result. What the rat is doing cannot be defined in terms of producing a sequence of movements, for as soon as a certain sequence is no longer capable of accomplishing the appropriate result (depressing the lever), the rat creatively discards it. Thus, there must be something going on inside the organism to accomplish the lever press under the various conditions defined by a given preparation. Skinner does not take an interest in the question of what this something is. He assigns it to the physiologist instead. But he is interested in defining exactly what the physiologist must explain. The relevant response is exactly the one specified by the experimental analysis of the unit of behavior. A sufficient basis for defining an elementary unit is for its members to acquire and lose their causal properties in a lawful manner. Skinner's attitude is: If the most elegant laws of learning attach to reflexes that fit the S-R mold, then so be it; but if we can formulate more elegant laws by violating the narrow constraints of S-R psychology, then violate the constraints.

The Active Organism.[4] Taking correlations of generic classes as

---

[4]Hilgard (1956) notes that there is "a family resemblance between Dewey's position and operant behavior, in which responses are coordinated with the stimuli to which they lead" (p. 329). Emphasis upon the active organism and the need for a non-elementalism in the analysis of stimulus and response provide additional examples of this family resemblance.

elementary is equivalent to setting certain questions aside. If we cannot define these units physiologically, then the fact that the organism can produce a generic response to a generic stimulus constitutes a rather advanced accomplishment on the part of the organism. Clearly, both the ability to discriminate the stimulus and the ability to perform the response require explanation. Skinner however is willing to let such accomplishments constitute brute facts of a behavior analysis.

Consider the following example of the type of generic stimulus a behavior analysis will routinely posit as part of an elementary unit of behavior. Suppose we project a polka dot ball through the wall of a pigeon's chamber just before we deliver a shock. When we project a plain white ball through the wall, no shock is delivered. If we do this over and over again, the polka dot ball's presence eventually elicits a conditioned response--say, head raising. What is the stimulus? On each trial, the animal interacts with the ball in different ways. It turns its head back and forth to change the angle of incident light, it moves towards or away from the object, perhaps even pecks it. It may well be impossible to define a class of receptor impulses that correspond to the presence of the polka dot ball. The stimulus identified as *projecting a polka dot ball into the chamber* can include episodes of complex interactions with the environment. The animal may not simply be passively receiving stimulation, but may be doing things to influence the sample. So if the experiment reveals presence-of-polka-dot-ball to be the generic stimulus, the fact that the presence of this object functions as a category in the animal's behavior may represent an advanced accomplishment on the animal's part. It may not simply be the activation of certain transducers that constitutes the stimulus, but the sequence of receptor impulses in relation to the sequence of movements. Or there may be even more complex processes at work. One does not know until one investigates. Meanwhile, a behavior analysis can proceed without needing to make any assumptions about the outcome of the investigation.

Similar considerations apply to the concept of the response. A lever press is actually a complex interactive process between the animal and its environment. Each such response may in fact be the result of a unique sequence of muscle contractions. A closer look might further reveal that the sequence of contractions is being adjusted in mid-sequence on the basis of sensory stimulation. Thus, what may from one point of view be termed a simple response is from another point of view a complex interactive behavioral process. Perhaps not all responses are like this, but some are. The point is that there may be a great deal of what an S-R psychologist would call complex behavior going on as part of the elementary stimulus and response of a behavior analysis.

The ability of the organism to respond appropriately to an indefinitely wide array of different receptor impulses is known as a *constancy*. One of the central problems of psychology is to explain how the organism manages to create a constancy. The generic concept of the stimulus takes such a constancy as a brute fact about behavior. The pigeon responds differentially to the presence of a polka dot ball. The polka dot ball is the stimulus. Identifying it as such does not solve the problem of how the indefinitely many different receptor impulses produced by the ball have the same effect on the organism. Identifying the stimulus generically simply brackets the problem. Ultimately, there must be an explanation of the ability of the organism to treat the receptor impulses produced by the ball in one way, while treating physically similar impulses not produced by the ball another way. One does not need to solve this problem, however, in order to formulate and progressively refine valid principles of behavior. So at least Skinner claims.

Similar comments apply to the ability of the organism to produce an indefinitely large number of different colorless movements that nonetheless are effective in achieving some constant result (moving the lever, depressing the key). This ability is sometimes described as a form of creativity. Again, this creativity constitutes a problem calling for explanation. Describing the lever pressing response generically does not solve this problem, it simply brackets it. Ultimately, there must be an explanation of the ability of the organism to produce exactly the right colorless movements that depress the lever, but not to produce those that miss the bar, raise it instead of lowering it, or fail to depress it with sufficient force to close the circuit. Fortunately, one does not need to solve this problem, however, in order to formulate and refine valid principles of behavior--claims Skinner.

The preceding points about the active organism are not limited to operant behavior. Robert Rescorla (1975), one of the leading contemporary theorists of respondent conditioning, makes a related observation. He points out an inverse relation between the complexity of one's behavioral categories and the complexity of the causal principles that connect them. If one begins with very simple units--the activation of transducers, the contraction of muscles--then one will need to refer to very complex causal principles to relate them to one another. On the other hand, if one begins with complex units--the presence-of-a-polka-dot-ball, the pressing of a lever--then one will discover relatively simple causal principles connecting them. Rescorla implies that even the much maligned process of association might be defensible if, instead of assuming that the end terms are physiologically defined stimuli and responses, we permit them to be generic stimuli and generic responses. The more successful programs of behavioral research have, it seems, taken the latter course, including research on both operant and respondent behavior.

An immediate implication of the preceding discussion is that operant psychology is not quite the right term to denote the program of research that has evolved out of Skinner's work. Although the program's most distinctive accomplishments may lie in the domain of operant behavior, its scope and ambitions are broader than that. Hence, a more inclusive term is needed, and the one that has come to serve this purpose is *behavior analysis*. So from this point forward, we shall follow standard practice, and speak of the *behavior analytic* program of research.

Stimuli Without Stimulation, Responses Without Movement. The depth of the chasm separating behavior analysis from S-R psychology is far greater than it appears, because Skinner's own physicalistic biases sometimes paper over the differences. But in fact there is nothing in the technique of behavior analysis that even implies a stimulus requires sensory stimulation. Skinner himself clearly assumes this is how the organism maintains contact with the environment. And certainly the obvious behavioral regularities are of this type. But nothing in the technique of behavior analysis would be jeopardized if it turned out that some organisms have a form of extra-sensory perception making it possible for them to adjust their behavior to events with which they have no sensory contact. Indeed, the best possible evidence of such extra-sensory perception would just be that a behavior analysis reveals a certain type of response to be a function of events with which the organism has no sensory contact.

Thus, suppose a rat named Carl is in a standard experimental chamber. The chamber is set up so that lever presses deliver food only when a certain switch is closed by the laboratory technician. When the switch is closed, every 40th press earns food. When it is not closed, presses are completely ineffective. The switch however is located in another room completely out of sight of Carl, but fully in sight of his identical twin, Clark. Suppose the technician opens and closes the switch in accordance with a random pattern. When the switch is closed, a bright light just next to it comes on until the switch is opened again. Now it is well known that if such a light were located inside Carl's chamber, he would press rapidly when it is on, but hardly at all when it is off. But Carl does not have such a light. Clark, on the other hand, sees a light that conveys the same information. Now suppose Carl's rate of lever pressing fits exactly the pattern it would if the light were in his chamber. This would constitute evidence of telepathic communication. There is nothing in such a result that contradicts or conflicts with the method of behavior analysis. It would be a fairly straightforward matter to add this regularity to an analysis of behavior, even though we would not have the first clue about how the twins accomplished such a remarkable feat.

The same point applies to responses. The obvious examples of responses are instances of the organism moving parts of its body. But the behavior analytic program is set up in such a way that evidence of psycho-kinesis (the ability to move objects without the use of bodily movements) could readily be gathered and incorporated into the analysis. Suppose, for example, a scientist were to put two experimental chambers next to one another with a glass partition separating them, so the animal in one chamber is able to see the animal in the other. The scientist arranges for the lever in one cage to operate the food dispenser in the adjoining one, and vice-versa. In a standard experiment, there is an animal in each cage. If they cooperate, they can feed one another. Suppose Carl and Clark are two such rats, and they are star performers in this sort of experiment, earning for each other a full ration of food on a daily basis. But one day, unbeknownst to the experimenters, Clark falls ill and lies motionless at the corner of his cage in full view of Carl. Then something remarkable happens. Without the benefit of any visible cause, the lever in Clark's chamber starts to pump up and down, and Carl gets his full ration of food as usual. When the scientist checks the cage and sees the limp Clark, and then looks at the recording device and realizes that Carl got his full ration, she would be on the brink of an amazing discovery. Carl can move Clark's lever without pressing it--a fact which she could confirm by showing that the rate of lever movements changes in an orderly manner as a function of (say) the number of movements required to produce a pellet in Carl's chamber. Here again, the result not only would not conflict with behavior analysis, it actually would be based upon evidence gathered from such sources. The best evidence that it is Carl who is causing the lever to move would be the fact that changes in the rate of lever movements fit the pattern we would expect of him.

Let's take our fantasy a step farther. Although it would be much more difficult to do, one could even establish experimentally that a rat could perform a psycho-kinetic response to a telepathically received stimulus. Suppose for example that Carl has a distinctive extinction curve: when a certain response no longer earns food, Carl performs a series of bursts of responses which bursts come in triads fitting a long-short-long...long-short-long pattern. No other rat has ever had an extinction curve quite like it. Now suppose Carl's identical twin, Clark, is placed in the paired chamber again, but this time without a partner. Meanwhile, Carl is moved to a laboratory across town. Now suppose the standard cooperative contingency between lever presses in one chamber and delivery of food in the neighboring chamber is put into effect, and the lever in the chamber next to Clark starts pumping up and down, and Clark eats about half a day's ration of food. Then suppose the contingency between lever pressing and delivery of food is halted. Clark no longer receives food deliveries, a situation that would normally evoke an extinction response from Clark if he himself were the one pressing the lever. And indeed, the lever in the adjoining cage does begin moving in fitful bursts. The pattern is the long-short-long...long-short-long extinction curve characteristic of Carl. At this point, the method of behavior analysis would force us to add another operant to Carl's repertoire. This operant would correlate a response that is not a physical movement with a controlling variable that is not a physical stimulus. Although I do not expect such a unit to be discovered, there is nothing in the method to prevent it. And if the experimental facts were as I have imagined them, it would be impossible for a behavior analyst to deny that such a unit is behaviorally real.

The point is that as conceived by behavior analysis, the field of behavior cannot be defined as (say) the bodily movements of an organism in response to sensory stimulation, because we do not know where the experimental method will lead us. True, what we start out with is a certain intuitively understood regularity between responses that require bodily movements and stimuli that require sensory stimulation. But this does not mean that in the process of refining our understanding of these regularities and exploring them in greater depth, we will not discover that the end terms can be quite different than we thought they could be. Once we have refined our understanding of regularities among familiar entities, we can use these regularities themselves to discover entities of an unforeseen nature. In this sense we discover what behavior is, rather than starting out with a definition of it that sets the limits of our scientific enterprise (as S-R psychology and common sense each in its own way would do).

Molar Behaviorism. If Skinner broke with S-R psychologists on the question of defining the units of behavior, he was not alone in this respect. As Verplanck (1954) says,

> Skinner wants to start with a point-at-able world, with point-at-able operations, and to carry on from there. He accepts as his point of departure the world of things and activities and leaves to others, who start reductively, the fields of "perception" and "sensation." It is often with surprise that persons most familiar with earlier frames of reference in psychology recognize that this is true of other current behaviorists. Physiological elementalism, in the style of Watson, is not a necessary characteristic of today's behaviorists. (p. 308)

Indeed, in the 1940's and 1950's many psychologists called themselves molar behaviorists, just to draw a sharp distinction between S-R psychology and a non-reductive alternative.

The term *molar behaviorism* was introduced by Tolman in the 1920's to refer to an approach to the study of behavior which assumes that behavioral regularities cannot be reduced to S-R bonds of physiologically defined end terms. Egon Brunswik (1952) uses this term to refer to a general strategy of behavioral psychology which searches out the "focal arcs" that relate "distal stimuli" to "distal responses." In his view, S-R psychology tried to relate "proximate causes" (the physiological stimulus) to "proximate effects" (muscle contractions), but was unsuccessful in finding regularities of this type. Molar behaviorism takes a different strategy. It defines the stimulus and response on the basis of the interaction of the organism with its environment. A stimulus can thus be an object of a certain description with which the organism has established (sensory) contact. The stimulus is "out there," and the analysis does not presume to show how the organism manages to discriminate its presence. And in this sense, the stimulus is distal rather than proximate. Likewise the response is defined in terms of the organism's effect upon the environment (pressing the lever, pecking the key, moving from the start of the maze to the goal box) instead of how the effect is achieved. Again, in this sense the response is distal rather than proximate.

Brunswik views an organism as a "stabilizer" that maintains a certain equilibrium between itself and its environment. To maintain effective contact with the environment is an accomplishment, both at the stimulus and the response side of the equation. The organism cannot always succeed in coming into sensory contact with its environment. Likewise, it does not always succeed in performing an effective response. Behavioral regularities therefore are, of necessity, probabilistic. The problem of analyzing behavior is to discover the points at which stimulus and response come into focus--i.e., at which the two come into the most significant causal relationship. Brunswik does not describe how one finds these focal points. But he includes Skinner among the focal arc theorists, and so far as I know, Skinner's method for defining stimulus and response is the earliest and most widely accepted treatment of this problem. Thus, Skinner would seem to be a prime example of a molar theorist.

Verplanck (1954) cautions, however, that "the cant terms 'molar' and 'molecular' cannot be applied in an intelligible way" to Skinner's position (p. 273), because under Skinner's system the behavior of the organism determines the level at which stimulus and response are defined. Thus, it is possible that the leg flexion reflex could be defined in strict molecular terms as a contraction of certain muscles in response to stimulation of certain transducers. Even the lever-press could in theory be molecular. Suppose that operation of the lever delivered food only when a certain set of muscles is employed. If one could get the most orderly dynamic laws by defining the lever-press in terms of these muscles, then Skinner would define the response in this instance at the molecular level. Clearly, he does not think this will happen in all cases. But he does maintain that the question of the level at which a unit of behavior is defined should be answered by the behavior of the organism itself. What he provides is a general method for extracting the answer, and a set of examples in which the experimentally determined level turns out to be molar.

CHAPTER THREE

THE FUNCTIONAL NATURE OF BEHAVIORAL CATEGORIES

To define each unit of behavior experimentally would be an enormous task. Even for an individual pigeon or rat it may not be possible because there are an indefinite number of such units in its behavior and new ones are always being formed. Behavior analysts do not however propose to study behavior unit by unit, animal by animal. The main objects of study are the principles by which units come into existence, undergo change, and go out of existence. Skinner assumes that each different kind of unit has its own such principles. His research strategy calls for intensive study of only a few carefully chosen representatives of each kind in a few representative species. If the dynamic regularities of a given kind of unit are consistent from one instance to another, and from one species to another, one may (in the absence of evidence to the contrary) reasonably proceed on the assumption that the regularities generalize to all units of the kind under study, for all species that possess the unit.

I

The first kinds of units to be studied were the respondent and the operant. Research on the respondent began with Pavlov's investigation of the dog's salivatory response, and came to include the rabbit's eye blink and the human being's galvanic skin response. Research on the operant began with Thorndike's investigation of the cat's latch-opening response, and came to include the rat's lever press, the pigeon's key peck, and the monkey's lever pull. As one might expect, the dynamic regularities of the operant and the respondent include constants that must be set for each species and individual. But the regularities themselves have proven remarkably general across individuals and species. The explanation of how this is possible requires us to explore the functional nature of behavioral terms.

The Ontology of Behavior. Behavior comes in individual units that maintain their identity through time, and these constitute the subject matter of behavior analysis. In order for an aspect of an organism's relationship with the world to constitute a behavioral unit, the environment and the organism must form a system. Such a system requires a causal relationship between part of the environment and part of behavior, but not every such relationship is part of a system. Suppose a sudden movement of Fido's leg is in response to a flea bite, so that the event of the flea biting the dog caused Fido's leg to twitch. Nothing in this insures that cause and effect are part of an organism/environment system. When environment and behavior form a system, there are causal regularities between them that extend over a considerable length of time. Perhaps a bite of the sort the flea just delivered to Fido causes a twitch of the sort Fido just performed only if 1001 different neuro-physiological factors are just right, and these factors all vary randomly. If so, then there is no sustained causal relationship between environment and behavior, and Fido's twitch in response to the flea bite does not instantiate a behavioral unit.

A bell rings and a hungry animal is presented with food. The stimulus of the food causes the animal to salivate. After repeated presentations of food in conjunction with the ringing of the bell, there is a change in the causal power of the ringing bell. The sound of the bell begins to have the power to cause the animal to salivate. It once was neutral, but it now has the ability to cause the animal to salivate. The sound of the bell, the presence of the food, and the response of salivating have formed a behavioral unit. We know such a unit exists because we can verify that certain aspects of the environment and certain aspects of behavior have an ongoing causal relationship with one another. By their very definition, behavioral units display such causal relationships. When such relationships come into existence, so do the units. And when they disappear, the units disappear also.

In the case of a respondent, the defining causal relationship goes from environment to behavior: a stimulus causes a response. In the case of an operant, the defining causal relationship goes in the opposite direction: a response causes a (reinforcing) stimulus. This stimulus in turn, however, has the power to alter future responding; so a response can alter the environment, which can alter responding, which can alter the environment, and so on. An animal wanders around the experimental chamber. Eventually, it presses a lever projecting from the wall of the chamber. Almost immediately, a pellet of food drops into a trough, and the animal eats it. The animal now returns to the lever and presses it again. Another pellet of food drops into the trough and the animal eats it. The stimulus provided by the food has made causal contact with the response of pressing the lever. An operant has come into existence. Lever presses cause the reinforcing stimulus of presentation of food, and this reinforcing stimulus raises the rate of lever pressing.

Respondents and operants are what a logical positivist would have called theoretical (as opposed to observation) terms. According to logical positivist epistemology, a causal relationship is always inferred nondeductively from empirical observations.[1] We observe the animal pressing the lever, then the delivery of food, then a higher rate of lever pressing. We cannot, however, observe that food delivery is causing the higher rate of lever pressing. But behavior analytic terms imply the existence of certain causal relations. One knows something is an operant response, discriminative stimulus, reinforcer, respondent response, unconditioned stimulus, or conditioned stimulus only if one knows its occurrence follows certain causal patterns. A response is operant only if it occurs more frequently as a result of being reinforced, a stimulus is a reinforcer only if it increases the rate of operant responding, and a stimulus is discriminative only if it controls the rate of responding as a result of its correlation with an operant/reinforcer contingency. A response is respondent only if it can be elicited by an unconditioned stimulus, a stimulus is unconditioned only if it can elicit a respondent without benefit of conditioning, and a

[1]There are now alternative philosophical analyses of observation, however, which incorporate inductive inference into the very act of seeing. One of the first of these was Dretske (1969).

stimulus is conditioned only if it is capable of eliciting a respondent as a result of its correlation with the occurrence of an unconditioned stimulus.

The concepts of behavior analysis are defined on the basis of an event's ability to cause certain types of effects, or to be the effect of certain types of causes. Such concepts are commonly referred to as *functional* concepts. For example, a certain arrangement of metal springs and wood is called a mousetrap, because it has the ability to cause mice to become trapped in it. A certain type of electrical device is called a metal detector, because it has components which are able to respond in a certain way to the presence of metal. The six central concepts of behavior analysis are the mousetraps and metal detectors of behavior. We apply these concepts to parts of the organism/environment system that function in certain ways. A stimulus that has the power to elicit responses is an unconditioned stimulus, and the responses it elicits are respondent responses. A stimulus that has the power to increase the frequency of responses that are capable of causing that stimulus is a reinforcer, and the responses it reinforces are operant responses. Stimuli that gain control over respondent responses by being correlated with unconditioned stimuli are conditioned stimuli. Stimuli that gain control over operant responses by being correlated with the existence of a contingency between responding and the delivery of reinforcement are discriminative stimuli. Like mouse traps and metal detectors, all six of these basic concepts are defined in terms of their causal roles.[2]

To establish that some aspect of the natural world fits a functionally defined category, we may need to study it carefully under controlled conditions for a considerable length of time. We want to know if puffs of warm air can function as reinforcers for lever pressing. So we

[2]The term *function/functional* is here taking on a different sense than it has in the phrase *functional analysis*. A functional analysis is an attempt to state an empirically valid quantitative relationship between two or more physically defined variables--e.g., the volume of a gas is proportional to its temperature and inversely proportional to its pressure. A functionally defined entity, on the other hand, is something that has certain causal properties by its very definition. It is unfortunate that the word *function*, has come to have these two distinct meanings, but it is too late to do anything about it.

establish a base line rate of lever pressing in the absence of any connection between lever pressing and puffs of air. Then we present a puff of air for each lever press, and note that the rate of lever pressing increases. Finally, we sever the connection between lever pressing and puffs of air, and find that the rate of pressing returns to its base line level. We have thereby discovered a behavioral unit, consisting of an operant response (lever presses) that is under the control of a reinforcer (puffs of warm air)--i.e., we have discovered that lever presses function as a response-for-warm-puffs-of-air and that warm puffs of air function as a reinforcer-for-lever-presses.

Not Offering Functional Explanations. The word function has so many different meanings that its use invites misunderstandings. Nonetheless, it is a word that has come to be used by behavior analysts, so we cannot ignore it. But we should take whatever steps are necessary to avoid predictable confusions.

A case in point is the question of whether functional concepts have a special explanatory force. Biologists sometimes attribute functions to parts of organisms: the function of the heart is to pump blood, the function of the kidney is to filter out impurities, and so on. In addition to pointing out something about the roles that the heart and the kidney play in the operation of the organism, such functional attributions are sometimes alleged to explain why the organs in question exist: hearts exist in order to pump blood, kidneys exist in order to filter impurities out of the blood. This interpretation then raises the question of whether such explanations attribute purposes to natural processes and thus explain natural events teleologically. Those who answer in the negative usually try to offer some other basis for the alleged explanatory force of functional attributions, the favorite candidate being the process of natural selection.

I do not wish to raise the question of whether functional attributions in biology have explanatory force. I mention this issue only to caution against an analogous interpretation of the functional categories of behavior analysis. To say, for example, that the delivery of food functions as a reinforcer is to say only that it increases the rate of some response upon which it is contingent. It is not, however, a functional attribution of the sort illustrated above, and therefore not a functional explanation of the kind that may or may not exist in biology. A behavior analysis does not, for example, imply that food is being delivered in order to increase the rate of a certain response (even though that will be the effect, if food delivery functions as a reinforcer). So-called functional categories play a purely descriptive role in behavior analysis. Applying them to the organism/environmental system is a prelude to a special type of causal explanation, but not itself an explanation--and definitely not a non-causal explanation.[3]

Categories That are Not Physically Definable. As many philosophers have pointed out, mouse traps and metal detectors can take indefinitely many different physical forms. This is because what qualifies something to be classified in one of these ways is its causal properties. And there are indefinitely many different physical arrangements which can produce such causal properties. An immediate corollary is that there is no way to define the property of being a mousetrap or metal detector in first-order physical terms. The key qualification for being a mousetrap is the ability to entrap mice, and there are so many different ways to do this that the set of mousetraps cannot be defined by some shared physical property or mechanism. Each mousetrap is a physical object, but what it has in common with the rest of the category is not a physical property (no matter how complex), but a functional one. The same is true of metal detectors. And the same is true of the categories that enter into the definition of the units of behavior--reinforcer, operant response, discriminative stimulus, unconditioned stimulus, respondent response, conditioned stimulus. What makes some aspect of the organism/environment system an instance of one of these categories is the causal role it plays--i.e., its function. So it follows that there is no physical property shared by all-and-only the instances of a given behavioral category--e.g., no sequence of muscle contractions and no topography that characterizes the category of lever presses.

When Skinner first drew the operant/respondent distinction, he emphasized that a type of response is classified as operant or

[3]See Rachlin (1992) for an unorthodox dissent from these cautionary remarks.

respondent based on whether it follows operant or respondent principles. A child's secretion of tears after a painful injury is a respondent, but the child who has learned to cry real tears upon receiving a reprimand because tears have been followed by reinforcing events (hugs, attentiveness, relaxation of rules, etc.) is emitting an operant (Skinner, 1937, p. 383). The responses themselves may be physically indistinguishable--indeed it is this similarity that makes operant tears so effective--but they are classified differently because of the different ways they relate to behavior causally.

The same is true of the classification of stimuli. A certain scent may sometimes cause an animal to approach but at other times cause it to withdraw. The scent would then be classified as two different stimuli, depending upon whether it functions as an attractant or as a repellent. Verplanck remarks upon this peculiar sense of the term stimulus.

> Here, "stimulus" refers to a class of environmental events that cannot be identified independently of observations of a specified activity of the organism and that must control that activity according to a specified set of laws. A red triangle of specified physical characteristics may not be termed a stimulus when it is repeatedly presented in association with food to, say, a dog, until the dog comes to salivate regularly in response to it according to the laws of behavior. Thereafter, it need be specified only insofar as it can be seen to control the *specified* behavior systematically. But if we find, upon further experimentation, that *any* red object controls the response, according to precisely the same laws, and also that triangles that are *not* red do not, then, by this usage, the red triangle can no longer be termed the stimulus, and something else, presumably "anything red," is the stimulus. And so, although we may empirically identify manipulable objects and events that we may call stimuli, we do this on the basis of a construct, in Skinner's case, the reflex--and the term stimulus is stripped of all data-language status. It is a quasi-independent variable, and when the term is used rigorously must be carefully stated as a stimulus-for-knee-jerk, stimulus-for-bar-press, and so on. (p. 285)

This approach to defining basic concepts is just the sort of thing positivists in the Dartmouth seminar objected to.[4]

Defined in terms of physically distinct movements, the number of different responses by which the rat presses the lever is "indefinite and very large" (Skinner, 1935a, p. 351). One press is accomplished from a sitting position, another as part of a running movement. The force for one press comes mainly from the left forepaw, for another from the right. Furthermore, if the location or size or torsion of the lever is changed, the physical movements of the animal will change accordingly. Nonetheless the rate at which the rat presses the lever continues to be a lawful function of certain third variables. The resulting changes in rate are so smooth that even though the responses at one time may be dissimilar to the responses at some other time, they are "quantitatively mutually replaceable" (p. 351).

This fact justifies treating a class of physical movements that cannot be defined physically--there is no set of movements in space that all-and-only lever pressing movements have in common--as a unitary response class. Thus, despite the fact that there are perhaps literally an infinite number of different sequences of muscle movements by which a rat may press a lever, we can treat them as part of the same reflex if they function in this quantitatively equivalent way within the animal's stream of behavior. The result is a response class that cannot

---

[4]Verplanck (1954) as usual notices exactly what Skinner is doing and summarizes it clearly and economically.

> Rather than being a set of empirical laws embodying statements that represent inductive generalizations based on a set of terms initially defined in a data language, it is a set of formally defined terms, and defining laws, which are only coordinated with data-language statements after they have been fully stated. Stimuli and responses cannot be identified independently of the theory; they are defined by the theory for the theory. Similarly, the central variable of the system, with which the experimental program has been preoccupied, the operation "reinforcement," rather than being inductively achieved as a central principle for the explanation of behavior, turns out to be a principle that serves, with some others, to define the area with which the theory deals. The actual independent variables of the system are different both from those of other systems and from those stated for the system. (p. 295)

be defined in terms of the topography of the animal's movements. There is no first order physical property that all-and-only lever presses have in common. This category can be defined only in terms of the animal's effect upon its environment.

In a similar manner, a stimulus is a reinforcer only if it causes the rate of responding to increase, but a given stimulus will function in this way only part of the time, and these instances have no physical property in common. Thus, there is no physical definition of a given reinforcing stimulus. Likewise, a stimulus is discriminative only insofar as it functions in a certain way in the animal's behavior, and the instances of the stimulus having this function in common cannot be defined physically.[5]

II

One of the longstanding criticisms of behavioral psychology is that it ignores differences between species. How can the same causal regularities apply to rats, pigeons, monkeys, and human beings? It seems absurd to think that animals so different in so many obvious respects could be identical in their patterns of behavior. Yet a surprisingly strong case can be made for such a claim, so long as behavioral terms are understood correctly. Once we understand the effect that the functional meaning of key behavioral concepts has upon the content of behavioral principles, it becomes plausible to think that they may have cross-species validity.

---

[5]Interestingly, Verplanck (1954) notes that Tinbergen's ethological concept of the releaser parallels the concept of stimulation used by Skinner--i.e., it too is defined functionally. Verplanck's implication that ethological analyses are compatible with behavior analyses proved prescient. Skinner acknowledges the compatibility by including released behavior alongside operant and respondent behavior as basic behavioral units (Skinner, 1981). And there is now a journal, *Behavioural Processes*, that melds behavior analysis and ethology into a single subject.

The Content of Behavioral Principles. Consider an idealized reconstruction of Skinner's investigation of the rat's lever press. First he discovered that in a certain type of environment (an experimental chamber) a certain type of response (the lever press) increases in rate when a certain type of stimulus (food delivery) is contingent upon the response. He took considerable care to adjust features of the environment--the lever, the food, the method for delivering it, the degree of food deprivation of the rat, and so on (using certain broad reliable secondary processes such as increased responding when the contingency is in effect, and a decline in rate after the contingency is extinguished)--so that the rate at which the rat pressed the lever was under the control of the delivery of food. He had arrived at a preparation in which the rat's lever press was an operant response and food delivery was a reinforcer.

At this point, he began to gather new information by altering third variables and observing the effect upon the rate of responding. For example, instead of delivering food after each response, one can deliver it only after three responses--a fixed-ratio 3 (FR 3) schedule of reinforcement. The animal is kept at this schedule until its behavior stabilizes into a regular pattern, and then held there for a considerable length of time just to confirm that stabilization has occurred. Such an experimental session lasts until the rat earns its day's ration of food or until a certain time limit (of perhaps 2 or 3 hours) is reached, whichever comes first. (If the time limit is reached first, the animal is given the rest of its ration later in the day.) A record is kept of the rate of response under this schedule, and compared with the rate under the earlier schedule. Subsequently the ratio can be stretched to require more responses in order to cause delivery of food. Each increase in ratio is accompanied by a higher rate of responding, although the rate of increase slows as the ratio gets higher. As one continues to stretch out the ratio, eventually a point is reached at which the animal simply quits responding. Prior to then, however, certain other features of responding emerge. Perhaps the most pronounced of these is the following distinctive pattern: the animal responds rapidly up to the point of reinforcement, then pauses after consuming the food, then rather abruptly begins a run of rapid responding up to the point of reinforcement again. When this pattern is graphed on a cumulative record, where the vertical axis represents the total number of responses and the horizontal axis represents time, the result resembles the profile of stair steps. This pattern constitutes a causal regularity in the operant behavior of the rat: Fixed-ratio schedules result in cumulative records having a stair step appearance.

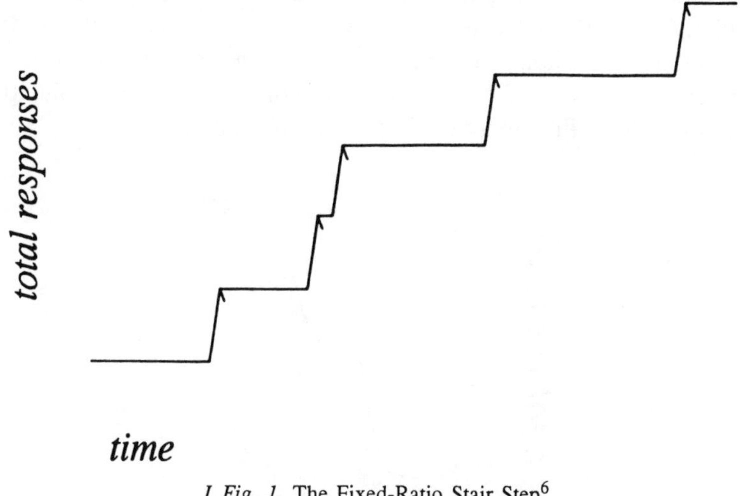

I Fig. 1. The Fixed-Ratio Stair Step[6]
(after Ferster & Skinner, 1957, p. 51)

Skinner also programmed his equipment so that a certain fixed-interval of time had to pass before lever pressing became effective. Once this interval terminated, the first press thereafter would result in food delivery. One can start with a relatively short duration--say 15 seconds. Again, the animal is kept on this schedule until its behavior stabilizes into a regular pattern, and then held there for a considerable length of time. The session lasts until a full day's ration of food has been earned or until a time limit is reached. A cumulative record can be kept of the changing rate of responding. Later, the interval can be lengthened and a record kept of the results, and then lengthened even further, and so on. Again, a pattern emerges. Once the animal has

---

[6]The five oblique marks at the outer edge of each step in Figure 1 indicate the times at which reinforcement has occurred.

had some experience with the day's schedule, its rate of responding changes in a regular way. At the beginning of the fixed-interval, there is hardly any responding at all, then about half way into the interval, a slow rate of responding begins, and the rate accelerates steadily until the interval terminates, and food is delivered. The rat eats the food. And a new interval begins. When this pattern is graphed on a cumulative record, the result looks like an ascending series of scallops. This pattern constitutes a second causal regularity in the operant behavior of the rat: Fixed-interval schedules result in cumulative records that look like an ascending series of scallops.

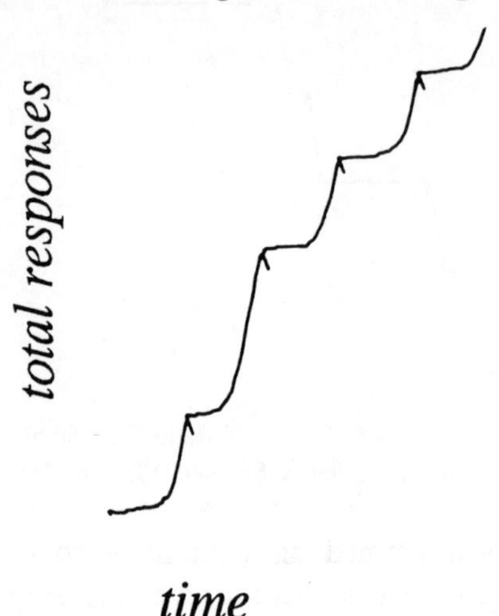

II Fig. 2. The Fixed-Interval Scallop
(after Ferster & Skinner, 1957, p. 159)

Skinner next tried switching back and forth between the two types of schedule. To give the rat an indication of what is happening, one can install lights in the chamber. When a fixed-ratio schedule is in effect, one light is on; when a fixed-interval schedule is in effect, the other light is on. One takes as many days as necessary to establish this discrimination. The animal now produces a clear stair-step pattern during the fixed-ratio portion of the session, and a crisp scallop pattern during the fixed-interval part of the session. What does this say about the behavior of the rat? One way of describing it would consist of a purely physical description of what occurred: When food was delivered on a fixed-ratio schedule in the presence of a certain light, the rate at which the circuit connected to the lever got closed followed a stair-step pattern. When food was delivered on a fixed-interval schedule in the presence of a different light, the rate at which the circuit got closed followed an ascending scallop pattern. This (so far as it goes) is an accurate description of what happened. The experimental protocol would permit us to include considerable detail about the actual sequence of physical events. We could say how high the lever was located, how much force was required to close the circuit on each press, how much food was delivered with each fulfillment of the contingency, how much time was allotted for the animal to consume its food pellet before resumption of the schedule, how this eating period was marked (e.g., by dowsing the house light and turning on the food dispenser light), what the physical parameters of the fixed-ratio and fixed-interval schedules were, where exactly lights were located, what their wattage was, and so on. Little of this, however, is included in the description of the experiment in so far as it is intended to represent a projectible behavioral pattern.

What is actually projected is based upon a functional description of the experiment--roughly, that when a rat alternates between two discriminated operants, one of which is on a fixed-ratio schedule and the other of which is on a fixed-interval schedule, the transition between schedules is marked by a virtually instantaneous shift from the stair-step pattern characteristic of the fixed-ratio schedule to the scallop pattern characteristic of the fixed-interval schedule. A great deal is implicit in such a description. It is assumed, for example, that one already has attained a focused unit of behavior--i.e., that one's preparation has achieved smooth secondary processes at the given specifications of physical parameters, including those associated with the discriminative stimuli. The schedule parameters are assumed to have made contact with the organism's behavior, the lights are assumed to be exercising control over rate of responding, and of course the delivery of food is assumed to be functioning as a reinforcer and the pressing of the lever is assumed to be functioning as an operant response. There are various other assumptions that are normally unstated, such as that the animal is at 80% of its free feeding weight, that it is a healthy representative of its genetic strain, etc.

Cross-Species Principles. The principle being projected from the experiment is conditional. It applies if, and only if, a certain type of behavioral regularity already exists. Certain aspects of the environment and of the animal's behavior must be functioning in a certain way. Given that the key functional categories apply to the animal's current behavior, the principle states how certain aspects of the environment (in this case, discriminative stimuli marking contingencies of reinforcement) control certain aspects of responding (in this case, rate).

Suppose we try to extend such a principle to another species. Obviously, we will not expect the same physical protocol to be in operation. A pigeon, for example, is not likely to press a lever or to be reinforced by delivery of pellets of rat food. So if we are to test the principle on pigeons, some adjustments must be made. The adjustments are actually rather extensive, although they are seldom discussed. The goal of the adjustments however is clear: to arrive at a combination of behavioral and environmental factors that function for the pigeon in the same way that the lever press, the delivery of food pellets, and the on/off condition of stationary lights function for the rat--i.e., to arrive at a focused preparation that results in a discriminated operant. This is a problem Skinner worked on during the early 1940's. As the functional equivalent of the rat's lever press he developed the pigeon's peck of a key, a plastic disc similar to those one presses in automatic elevators. Instead of delivering pellets of food, he gave the pigeon a few seconds of access to a hopper containing bird seed. And instead of turning lights on and off in various locations, he took advantage of the pigeon's ability to discriminate colors by projecting different colored lights on the key. Then he experimented with various physical dimensions of these factors until he found a combination that came into focus. The key must be at a certain height, have a certain size, require a certain force. The hopper contains a certain mixture of seeds, is located in a certain relation to the key, is held in feeding position a certain length of time. The animal is given time to adjust to the sound of the hopper swinging into place and to the presence of a light above the hopper which indicates the duration of the feeding interval. The number of key pecks required to move the pen of the cumulative record up an inch is adjusted so that changes in rate make a visible difference in the record. (The pigeon can peck the key much more rapidly than a rat can press the lever.)

Having made all these adjustments so we have a functionally equivalent preparation for the pigeon, we now can test to see if we get the same regularities. And we do. For example, fixed-ratio schedules still result in the stair-step pattern. Fixed-interval schedules still result in the scallop pattern. And once we have established (say) red and green as discriminative stimuli for fixed-ratio and fixed-interval respectively, we can get the same sharp transition from stair-steps to scallops to stair-steps again that we got with the rat.

Skinner himself continued to work almost exclusively with the pigeon, but others went on to develop a preparation with the monkey in which the lever pull becomes the operant response. Again, the same basic patterns on fixed-ratio and fixed-interval schedules are observed. In the following passage, Skinner describes an outcome of the research on the rat, the pigeon, and the monkey.

> Figure 14 shows tracings of three curves which report behavior in response to a multiple fixed-interval fixed-ratio schedule. The hatches mark reinforcements. Separating them in some cases are short, steep lines showing a high constant rate on a fixed-ratio schedule and, in others, somewhat longer "scallops" showing a smooth acceleration as the organism shifts from a very low rate just after reinforcement to a higher rate at the end of the fixed interval. The values of the intervals and ratios, the states of deprivation, and the exposures to the schedules were different in the three cases, but except for these details the curves are quite similar. Now, one of them was made by a *pigeon* in some experiments by Ferster and me, one was made by a *rat* in an experiment on anoxia by Lohr, and the third was made by a *monkey* in Karl Pribram's laboratory at the Hartford Institute. Pigeon, rat, monkey, which is which? It doesn't matter. Of course, these three species have behavioral repertoires which are as different as their anatomies. But once your have allowed for differences in the ways in which they make contact with the environment, and in the ways in which they act upon the environment, what remains of their behavior shows astonishingly similar properties. Mice, cats, dogs, and human children could have added other curves to this figure. (Skinner, 1956, pp. 94-95)

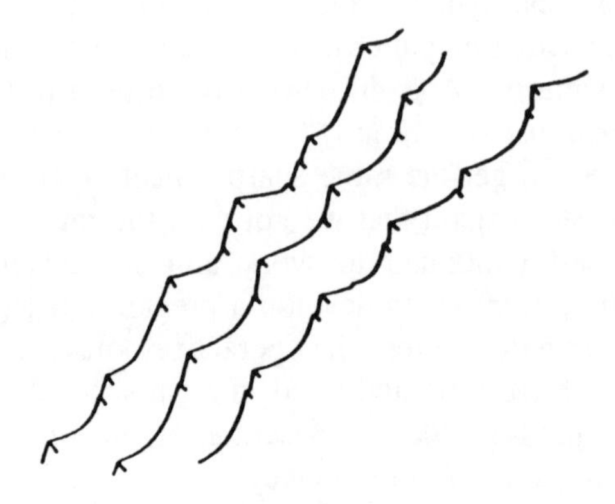

III Fig. 14
(after Skinner, 1956, p. 94, Fig. 14)

These cross-species regularities were not anticipated by any pre-experimental hypothesis of Skinner or anyone else.

Obviously, Skinner does not mean to imply that any regularity discovered for one species will be true of all others. This qualification is especially germane for regularities described outside the conceptual framework of behavior analysis. There is no implication, for example, that an arbitrary response can become an operant, that an arbitrary stimulus can become discriminative, or that an arbitrary biologically significant stimulus will function as a reinforcer. Although it would be very useful to have principles of this sort, they are not the sort a behavior analysis aims at. One might hope to be able to say which responses can become operants, but it is not likely one can do this validly across species. Likewise, one might aspire to say which stimuli can function discriminatively and which can function as reinforcers, but again it is virtually certain that few regularities of this sort will project validly across species. Thus, even though the experiments described by behavior analysts relate physical causes to physical effects, the regularities cut across environment-to-behavior relationships in an unusual way that is mediated by dissimilar mechanisms. Different organisms will have quite different operant repertoires. The aspects of the environment they find to be reinforcing will obviously be different. And the features of the environment which can function discriminatively for them will likewise be subject to enormous differences. Nonetheless it appears that once we take these differences into account, we can validly project behavioral regularities to a surprisingly wide range of species.

PART TWO

CIRCUMVENTING STANDARD CRITICISMS OF THE PROGRAM

> Controversies, such as those over "latent learning," and "continuity" and "discontinuity" interpretations are pointless within the Skinnerian framework. . . . To a remarkable degree, the theory is applied only to behavioral experimentation in its defined area, so that it "fails" to handle many data for the simple reason that it does not attempt to do so. (Verplanck, 1954, pp. 307, 309)

Skinner's terminology borrows so heavily from Pavlov, Sherrington, and Watson--i.e., from physiological reductionists--that behavior analysis appears to inherit all the shortcomings of a physiological approach to behavior. It takes a careful reading of Skinner to realize he gives terms such as *reflex*, *stimulus*, and *response* meanings quite at odds with their original physiological meaning. This has not only given his critics ample opportunity for misunderstanding, but has made it difficult even for Skinner's own students and co-workers to understand him.

The theme of the following chapters is that a number of traditional criticisms of behavioral psychology, sometimes thought to be decisive (either individually or collectively), nevertheless do not make contact with behavior analysis (correctly understood). The groundwork has been laid in Part I, where we explained why Skinner's terms do not always mean what they seem to mean. We shall now trace out some of the implications of that analysis.

CHAPTER FOUR

MINOR PROBLEMS

This chapter is devoted to what I am (perhaps somewhat misleadingly) calling minor problems. By calling them minor, I do not wish to imply they are insignificant (in which case they would not even merit discussion). I wish instead to distinguish them from an even more difficult problem to be discussed in succeeding chapters. Relative to that problem, these are minor. Nevertheless, each of the following lines of argumentation have had proponents who thought they offered sufficient grounds for dismissing behavior analysis as a deeply flawed scientific venture. And for this reason, these problems deserve our attention.

Behavior Analysis Ignores Biological Constraints. Behavioral psychology is often accused of extreme environmentalism, of ignoring all the interesting and rich differences between species. One species is able to form a discrimination of a certain property almost effortlessly whereas another is incapable of forming the discrimination even after extensive training. One species learns a certain operant response with minimal shaping whereas another cannot learn the response even with extensive training. The principles of behavior analysis may seem to gloss over these differences between species.

Such an accusation however misses the subtle way behavioral concepts finesse (as opposed to ignoring) species differences, in order to state generalizations that apply to animals that are in most other respects quite different. It is true that behavior analysis fails to focus upon many interesting differences between species. But differences in how animals operate upon the environment, what responses they can learn, how they gather information about the environment, how they process that information, and what stimuli reinforce their responses are irrelevant to the principle of the fixed-interval scallop or the principle of the fixed-ratio stair step. These principles successfully cut across such species differences. Just how far they extend is of course an open question. Biological research has often followed the strategy of studying some phenomenon intensively in one or two species, and then investigating other species to check the generality of the results. When problems are framed in a certain way, the results for one species are often quite general. And we have seen reason to believe that behavior analysis frames its problems in such a way as to achieve such generality. So biological constraints are not the decisive objection they are sometimes made out to be.

It Cannot Account for Creativity. Some critics have suggested that (strictly speaking) one could refute operant theory simply by changing the location of the lever in the operant chamber in the midst of an experiment. This would show that the animal is not the automaton that behavioral psychology supposedly assumes it to be, because when the lever is relocated, the animal will not press thin air but will adjust its movements to coincide with the new location of the lever.

Such a criticism misses what Skinner called the generic aspect of behavioral concepts. An operant is not defined as a sequence of muscle movements. Thus, operant theory does not imply that if the lever is moved, the animal will continue to perform the same sequence of muscle movements that resulted in reinforcement in the past--i.e., a sequence which now would have the animal pawing at empty space.

In any given environment there are likely to be an infinite number of ways for the animal to move its body through space to fulfill a given contingency: there are an infinite number of different sequences of muscle movements that would result in a lever press, for example. We have seen that behavior analysis does not address the question of what (if anything, other than the fact that they fulfill the contingency) these infinitely many different sequences have in common, nor does it address the question of what underlying process accounts for the fact that these are members of the same behavioral category. Therefore it does not address what, from a certain point of view, is the first question one might ask about operant behavior--viz., how does the organism coordinate its movements so as to bring about, through an indefinite number of distinct means, the effect that produces reinforcement? Instead, the behavior analyst simply notes that the organism is capable of functioning in a manner that fulfills the requirements of the contingency, and places all such responses within the same behavioral category--which is not to deny that there is an important question here, but simply to affirm that one does not have to address this question in order to study certain important aspects of behavior.

It is true that the animal shows a kind of creativity when it adjusts its sequence of muscle movements in a manner that is appropriate to the new location of the lever. Indeed, such creativity is also exhibited even if the lever has not been relocated. For--especially on lean schedules--the animal will roam around the chamber between lever presses, and therefore will often approach the lever from novel angles and postures. If one defines the dependent variable in terms of sequences of muscle movements, then operant theory has nothing interesting to say about the seemingly unbounded creativity of the white rat. But to interpret this fact as a critique of behavior analysis is to miss precisely the point that by defining the response class generically, one finesses questions about creativity and intelligence that can be raised in this manner. This does not mean the questions are answered. But it does mean that the behavior analyst is not responsible for answering them. (Although Skinner himself, at the peak of his career as an experimental psychologist, took on some additional responsibilities--as we shall see in our discussion of radical behaviorism.)

The concept of the discriminative stimulus is subject to a similar misunderstanding. To say that a certain class of events can function as a discriminative stimulus for a given animal is not to say what (if any) physical property these events have in common. It also is not to say what process accounts for the ability of the organism to discriminate these events from all others. The face of an individual human being serves as an effective discriminative stimulus for human beings, even though it is very difficult to say what all the different appearances of a given face have in common that distinguish them from the appearances of all other faces. And even if we knew what the appearances have in common, we would still not know how people are able to perform the discrimination, or why they are able to accomplish this feat with relative ease and yet must struggle to learn how to discriminate one species of sparrow from another.

Behavior Analysis is Incompatible with Cognitive Psychology. Questions about how organisms discriminate complex properties, or how they perform complex responses, have become central questions of cognitive psychology, which is now well established as the leading source of answers to them. Many people count this against behavior analysis, on the grounds that the two approaches contradict one another. But there is no contradiction between a cognitive explanation of how a stimulus functions discriminatively and a behavior analysis of the consequences of such a function for the rate of responding.

Take a simple stimulus, such as the presence of a red light. The explanation of the organism's capacity to discriminate this stimulus from other stimuli is of little, if any, interest to behavior analysis. The question is referred to the physiologist or cognitive psychologist. There also is a question about how the presence of the light controls a certain operant response. Here we must distinguish between the question of what causes the stimulus to come to function in this way and the question of what underlies this function. The former can be answered with a description of the procedure that causes the red light to acquire a discriminative function (typically, exposure to a situation in which a given contingency of reinforcement is in effect when and only when the red light is on). The latter, however, requires a theory of the underlying processes or structures that this procedure brings into existence.

Prior to the era of cognitive science, the dominant answer to the second question was the S-R theory of association-and-induction. Exposure to a certain procedure was supposed to cause the formation of an association between the discriminative stimulus and the contingency of reinforcement. Then stimulus induction expanded the class of effective stimuli to include those similar to the stimulus associated with the contingency. There are still theorists who attempt to account for the effect of discriminative stimuli in this way, but cognitive theory provides many other possibilities. There is no reason for behavior analysis to takes sides or choose favorites in the scientific competition to answer this question.

There is no incompatibility between analyzing behavior into the three-term contingency of discriminative stimulus/operant response/reinforcing stimulus and giving a cognitive explanation of how something comes to function as discriminative stimulus or operant response or reinforcing stimulus. Tradition, for example, attributes the following anecdote to the ancient Stoic logicians.

> A dog is tracking a scent when it comes to a fork in the road. It takes the right hand fork, follows it for a short distance, but fails to pick up the scent. At this point, without bothering to return to the place where the road forked, it takes a direct route to the other path, joining it some distance beyond the fork.

The Stoics thought this sequence of events showed that even a dog can reason syllogistically, because it must be making the following inference.

> The scent follows either the left fork or the right fork.
> It does not follow the right fork.
> Therefore, it follows the left fork.

Having arrived at the conclusion that the scent follows the left fork, the animal has no need to retrace its steps to the fork in the road, but proceeds immediately to the other path.

This account, now more than 2,000 years old, can serve as a simple model of how a behavior analysis is compatible with a cognitive explanation of the processes underlying behavioral capacities. We can describe the key events behaviorally as follows. The act of switching from the right hand path to the left hand path is an operant response, the absence-of-the-scent is a discriminative stimulus, and the recovery-of-the-scent is a reinforcing stimulus. We could say that the absence-of-the-scent functioned as a discriminative stimulus for the contingency that movement-from-right-fork-to-left-fork will lead to recovery-of-the-scent. This is simply a redescription in behavioral terms. In itself, such a redescription does nothing to explain the response in question, because a behavioral explanation would address the question of how the rate of the operant response relates to various independent variables such as the contingency between responses and reinforcement, the length of delay between response and delivery of reinforcement, etc. The hypothetical cognitive process, on the other hand, could explain something--namely, how absence-of-the-scent was capable of functioning as a discriminative stimulus for the contingency that movement-from-right-to-left will lead to recovery-of-the-scent. To say that absence-of-the-scent was the discriminative stimulus that caused the dog to switch forks (and not, say, the absence of broken twigs on the right-hand path) is simply to tell us what aspect of the environment initiated the switch, but a theory of underlying processes can explain how that stimulus brought the switch about--i.e., the cognitive explanation tells us what sort of process underlies the discriminative function of the stimulus upon this occasion (cf. Dretske, 1988).

A somewhat different role for cognitive processes may arise in the case of analyzing certain complex discriminative capacities. Consider the standard situation posed for a linguistic informant. The linguist presents the informant with sequences of morphemes, and the informant is supposed to categorize them as grammatical or ungrammatical. If the informant is a competent speaker, then the property of being a grammatical sentence constitutes a discriminative stimulus for (say) an affirmative nod. How can this complex stimulus function in this way? In addition to explaining what process connects the stimulus with an affirmative nod (if I do a good job identifying grammatical sentences, maybe the linguist will help me get a scholarship), there is the question of how the informant discriminates between grammatical and ungrammatical stimuli. This is the question a theory of grammatical competence is supposed to answer. Saying that the speaker simply generalizes from previous experience to the new situation is not an answer to this question. Calling the basis for generalization the process of forming an analogy is vague. These are telling criticisms of the theory of association. But associationism is the theory of S-R psychology, not of behavior analysis (although we shall need to complicate this picture once we turn to the discussion of radical behaviorism, as opposed to behavior analysis).

Turning to the concept of the reinforcer, let us consider a certain class of stimuli that have functioned as reinforcers on past occasions--say access to food. Again, one may ask several different questions.

What do these stimuli have in common, how does the organism pick out these stimuli as members of the same class, and what underlies their reinforcing function? The first question is not often asked--perhaps because the sorts of reinforcers used in experimental procedures are usually relatively simple stimuli--food, water, electric shock. But in the world outside the laboratory, reinforcers come in an amazing variety of forms. Indeed, according to recent conceptions of reinforcement, it would be a mistake to conceive of reinforcers as a class of stimuli. Most operant psychologists now follow Premack (1965) in defining reinforcers relativistically. That is, instead of conceiving of being-a-reinforcer as a property that some things have and others do not, they conceive of reinforcement as a relationship that exists between some events and not others. Thus, a given event (say, access to a running wheel) might serve as a reinforcer for lever pressing (so that an animal would press a lever to gain access to a running wheel), but not as a reinforcer for eating a food pellet (so that an animal would not eat food pellets to gain access to a running wheel). On this relational view of reinforcement, the reinforcer relationship orders the organism's environment rather than dividing it into a pair of categories. There remains, of course, the question of how the organism accomplishes this ordering, but this is distinguishable from the question of what quantitative behavioral effects follow quantitative changes in the way reinforcement is delivered (which is an important question of behavior analysis).

It is true, however, that some behavior theorists have attempted to address the question of what underlies the reinforcing function of stimuli. The most famous such theory is perhaps the drive reduction account of reinforcement--a theory defended by Clark Hull. But this is the type of question a behavior analysis sidesteps. No doubt the reinforcing function of stimuli may sometimes occur as the result of very basic physiological processes (as when a subject finds an event reinforcing because it feels good). At other times it may occur as the result of complex cognitive processes (as when a subject finds an event reinforcing because the subject desires goal G, and believes that the occurrence of the event increases the probability of G). Behavior analysis takes reinforcement in these and other cases to be a projectible predicate. That is to say, it assumes that the results of experiments performed with (say) biologically significant reinforcers such as food or water can be generalized to situations involving other types of reinforcers.

So far this assumption has led to a productive program of research. As a working hypothesis, it has received strong support, despite our lack of knowledge about the many different processes that underlie reinforcement. Certainly there are many different types of reinforcing events: those that have an obvious biological function (delivery of food, water, etc.), those which provide information that a biologically significant event is about to occur (such as a tone that precedes delivery of food), those which provide information that a certain response is correct (as when a teacher nods her head affirmatively when a student gives the right answer), those which record progress towards a goal (as when a person who is trying to lose weight enters his weight each week on a chart), and many others. One often hears the criticism that some stimuli increase the rate of a certain response only because they inform the organism that it got the right answer or is on the way to meeting a certain goal. The implication is that such events are not really instances of reinforcement. But this is to assume that the concept of reinforcement is linked to a certain account of what underlies reinforcement, which it is not. One advantage of approaching the analysis of behavior by way of functional categories is to finesse the necessity to theorize about such questions. Of course some psychologists will find these questions to be of central importance. Obviously, a purely functional conception of reinforcement does not address their questions.

Science Requires Underlying Entities. Some critics believe that progress in science requires incorporation of underlying processes and mechanisms into one's theoretical account (e.g., Cohen, 1984; Sosa, 1984). This does not, however, seem always to be the case. Sociobiology entertains a number of theses about social behavior that apply regardless of the mechanisms that underlie the behavior (Sober, 1985a). In this particular case, the generalization is broader precisely because it fails to incorporate underlying processes. And Newton's theory of gravity made no attempt to specify the underlying mechanism by which the gravitational force of one object acted upon another.

Perhaps Cohen and Sosa are confusing the behavior analytic program with Skinner's radical behaviorist philosophy. It is true that behaviorism rejects a certain kind of underlying entity (viz., mental ones), and may as a result impede scientific progress by inhibiting the explanation of behavioral regularities on the basis of regularities at a more basic level. As we shall see, some contemporary behavior analysts reject behaviorist philosophy for exactly this reason. But rejection of behaviorism is not incompatible with the goal of attaining continued progress (including increased theoretical integration of experimentally derived generalizations) at the behavioral level.

The Tautology Problem. The law of effect states that the rate at which a response is performed will temporarily increase as a result of reinforcing it. A common example of a reinforcing stimulus is access to food. Pigeons feed on grain. If a pigeon's key pecks are followed by access to a hopper filled with grain, then the law of effect would seem to entail that pecking should increase in frequency--which it usually does, especially if the pigeon has not eaten in a while. But suppose the pigeon has just finished eating to the point of satiation, and it shows no interest in an open tray of grain. What does the law of effect say now? Does it falsely imply that the pigeon will peck the key more rapidly when pecking produces access to grain? And does the pigeon's failure to increase its rate of responding refute the law?

The usual answer to both questions is no. The implication that the pigeon will peck the key more rapidly follows only if we assume that access to food is reinforcing. And the fact that the pigeon has just had a full meal implies that access to food will not be reinforcing--not, at least, until the pigeon is hungry again. And how will we know when the pigeon is hungry? Well, as soon as access to grain can increase the frequency of responses upon which it is contingent. So if we leave the key connected to the hopper, then we can infer that the pigeon is hungry and access to grain is reinforcing just as soon as pecks increase in frequency when followed by access to grain.

In practice, then, a stimulus is classified as a reinforcer when (and only when) it can increase the frequency of responses upon which it is contingent. But then a response which is reinforced will by definition increase in frequency. For if there is no increase in frequency, we will not classify the stimulus as reinforcing. In brief, we seem to be using the concept of reinforcement in such a way as to make the law of effect a tautology. This usage renders the law of effect irrefutable. But the cost of irrefutability is loss of empirical content. We demote the law from empirical generalization to stipulative definition. And some critics claim that as a result operant theory fails to satisfy the standards of scientific adequacy. For its basic principle is alleged to be a mere tautology, and is thereby rendered incapable of explaining events that fall under it. This, in brief, is the tautology problem.

The conventional way to attack this problem is to attempt to insure that the law of effect is not a tautology by defining reinforcement independently of rate of responding. The most plausible such definition states that reinforcement occurs when and only when a physiologically defined basic drive is reduced. This definition has the merit of tackling the problem head on. So long as one can verify drive reduction independently of the rate of behavior, the supposed circularity of the law of effect is broken. Unfortunately, drive reduction in this sense does not always appear to accompany what most psychologists mean by reinforcement. For example, an animal will repeatedly press a lever if doing so will give it an opportunity to explore unfamiliar environment or to view other animals, but there is no known chemical entity that is replenished or reduced as a result of such explorations or viewings.[1] These and other examples seem to indicate that the drive reduction definition is false. This does not show it is impossible to give an independent characterization of reinforcement, of course, but whether any alternative would fare better than the drive reduction hypothesis is questionable, given that drive reduction is the most plausible candidate.

An approach to the tautology problem more in keeping with behavior analytic practice would be to concede that the law of effect is a tautology, but to argue that this is not a problem. Skinner in fact has explicitly defined his basic concepts in such a way as to render the law of effect tautological. But he treated circularity as a good thing, as something to be done intentionally, and even as something to extend to other areas of the theory. This constitutes a distinctively Skinnerian

---

[1]I owe this point to Richard Colker.

approach to the tautology problem. It is one of the more subtle aspects of operant theory, and to my knowledge, something unique to it. Let us see how it works.

Take a pigeon that has not been fed since yesterday, so that access to grain is likely to be reinforcing. Instead of connecting the hopper to a manipulandum in the chamber, suppose we connect the hopper to a hand-operated button that permits us to give the pigeon access to food whenever we see a target response occurring. And instead of reinforcing key pecks, suppose we try to reinforce preening. Just to get things started, we lodge some particles of sand in the pigeon's feathers. The pigeon preens, we press the button, the hopper rises to the feeding position, and the pigeon eats some grain. Soon the hopper recedes from reach, and the pigeon eventually returns to preening. We press the button, the pigeon eats, and so on. Suppose this continues until the pigeon removes all the grains of sand, but at this point, the pigeon stops preening. Now what do we say?

One approach would be to say that evidently access to food is not currently reinforcing. But if we were to test this thesis by making access to food contingent upon some other response such as raising the head above a certain point, we would find head raising to increase in frequency. What this means is that some responses can be reinforced by access to food, but not the response of preening. Evidently the problem is not with our reinforcer, but with our target response. By attempting to reinforce preening, we seem to have chosen a response to which the law of effect does not apply.

We already know, of course, how to save the law from refutation. We simply do for response what we did for reinforcement--i.e., we link it by definition to the law of effect. We could for example define a subcategory of responses called operant responses. For a response to be operant, it must be possible to increase the frequency of the response by means of reinforcement. We now restate the law of effect to say that if an operant response is reinforced it will increase in frequency. If preening cannot be controlled by contingencies of reinforcement, this does not refute the law of effect. It only shows that preening is not an operant. And we have now apparently insulated the law of effect from refutation along a second dimension.

An expanded version of the law of effect will permit us to add and insulate yet a third dimension. This version states that if a stimulus is correlated with the condition that a given operant response will be followed by reinforcement, then the frequency of that response will increase in the presence of that stimulus. This gives us an opportunity to broaden the circle of definitions to include not only operant response and reinforcer, but also discriminative stimulus. Returning to our pigeon, suppose we arrange for the key to be lit when, and only when, key pecking will result in access to food. Soon the pigeon comes to peck only when the key is lit. Now suppose we decide to train the animal to respond in the presence of a more abstract stimulus. We arrange to flash polygons on the key. When they are equilateral, key pecking will produce access to food. But when they are not equilateral, key pecking is ineffective. To our dismay, the pigeon behaves no differently in the presence of equilateral polygons than in the presence of non-equilateral ones, even after extensive training. Does this failure refute the expanded law of effect? Not necessarily.

By following the strategy introduced above, we can save the law by defining a subcategory of stimuli called discriminative stimuli. A discriminative stimulus, by definition, will be one which comes to control the rate of operant responding when it is correlated with a certain contingency of reinforcement. And the law of effect will now state that if a discriminative stimulus is correlated with the effectiveness of a given operant response in producing reinforcement, then the frequency of that response will tend to increase in the presence of the stimulus. So if a given stimulus (e.g., an equilateral, as opposed to an non-equilateral, polygon) is correlated with the availability of reinforcement, but is nonetheless ineffective in controlling the rate of operant responding, then by definition it is not a discriminative stimulus. Such stimuli do not refute the law of effect, they merely fall outside the domain of phenomena the law applies to.[2]

This way of extending the circularity of the law of effect fits the

---

[2]Further refinements of this approach are necessary if it is to be rendered consistent with current knowledge about conditioning. For example, instead of defining the relation between the discriminative stimulus and the contingency as the former being correlated with the latter, we need to define it as the former providing marginally better information about the presence of the latter.

practice of behavior analysts. Another way of describing this practice is to say that the concepts of reinforcer, operant response, and discriminative stimulus are defined functionally, by means of the law of effect. When stimuli or responses function in a certain way, they fit operant categories. When they do not function in these ways, they do not fit operant categories.[3] This classificatory strategy traces back to Skinner's earliest work. He found that by describing responses and stimuli in functional terms he could arrive at broad generalizations about behavior. But the law of effect is not an exemplar of such generalizations.

The content of operant psychology lies not in the law of effect, but in the quantitative analysis of the law of effect. By definition, the delivery of a reinforcer increases the rate of operant responses which it follows, but there is nothing in the definition to say how much the rate will increase or how rapidly this will occur or how long the increase will last. And the answers to these quantitative questions turn out to be given by aspects of the environment. The category of reinforcing stimulus may itself not be physically definable, but given that some event is reinforcing, certain physical properties of that event have lawful effects upon the behavior that is reinforced. For example, the amount of reinforcement delivered or the schedule by which reinforcement is delivered or the length of delay with which reinforcement is delivered all have a measurable effect upon the rate of operant responding. Furthermore, although the property of functioning as a reinforcer cannot be verified independently of the occurrence of the response that is being reinforced, these physical aspects of reinforcement (e.g., the length of delay of delivery of reinforcement) can be verified and measured independently of certain physical aspects of the operant response (e.g., the rate of operant responding). These properties are the independent and dependent variables of a behavior analysis (see Morse & Kelleher, 1977).

Cumulative records provide convenient representations of such quantitative relationships. For a specific experimental session these

---

[3]For a recent discussion of the thorough and explicit manner in which behavior analysts continue to define their basic concepts functionally via the law of effect, see Gewirtz and Pelaez-Nogueras (1992).

relationships can normally be given a strictly physical description (a red light signaled the contingency that the tenth key peck having 1.5 grams of force or greater would be followed immediately by a two second delay during which the key turned white, followed by four seconds of access to the magazine).[4] What functions as a discriminative stimulus, or as a reinforcer, or as an operant response for one organism may not function in this manner for another. Thus, the generalization that fixed-interval schedules generate records with inverted scallops does not mean that you can take an arbitrary animal, an arbitrary response, an arbitrary detectable stimulus, an arbitrary pleasant stimulus, and then deliver the pleasant stimulus in the presence of the detectable stimulus according to a fixed-interval schedule requiring the arbitrary response, and the animal's frequency of responding will generate a series of inverted scallops on the cumulative recorder. Rather, it means that if you take an animal for which a certain triad of stimulus-response-consequence function as discriminative stimulus, operant response, and reinforcer respectively, and you deliver the reinforcer in the presence of the discriminative stimulus according to a fixed-interval schedule requiring the operant response, then the animal's frequency of responding will generate a series of inverted scallops. The functional nature of the concepts give the generalization an implicit conditional nature. Such generalizations cut across the environment and behavior in unusual ways that defy physical or cognitive description.

The dependent and independent variables of behavioral psychology are not its functionally defined theoretical entities, but certain physical parameters of them (Morse & Kelleher, 1977). It is these parameters that can be measured and controlled, and that stand in lawful relations with one another. The length of the pause in the fixed-interval scallop is proportional to the length of the interval; the length of the pause in the fixed-ratio stair step is proportional to the magnitude of the ratio; the distribution of responses between two identical variable-interval schedules matches the ratio of the inverse of the delay to reinforcement

[4]There are exceptions. Herrnstein, Loveland & Cable (1976) discovered that some stimuli that function discriminatively for pigeons have no physical description--e.g., truck part or part of tree.

on each schedule. These causal relationships are not in any way equivalent to the law of effect, although they take the law of effect as their point of departure. "The empirical law of effect is not the final word, but the beginning, a way of *framing* the problems to be addressed" (Hinson, 1987, p. 188, italics in the original). Hence, its circularity is a virtue, because it makes possible a complex functional definition of the components of a system of behavior.

Similar comments apply to the concepts used in formulating classical (respondent) conditioning. Here again one finds a complex circular definition which creates a point of departure for the quantitative analysis of a specific form of behavior. In this case, the concepts are those of unconditioned stimulus, unconditioned response, conditioned stimulus, and conditioned response. If prior to any form of conditioning, the presentation of a given stimulus elicits a given response, then the stimulus is said to function as an unconditioned stimulus in relation to the unconditioned response. If pairing a previously neutral stimulus with the unconditioned stimulus causes the neutral stimulus to elicit some response, then the previously neutral stimulus is said to function as a conditioned stimulus and the response to it is said to function as a conditioned response.[5]

---

[5]Actually, such functional definitions, while circular, are not totally immune from empirical refutation--of a sort. What one wants are definitions that define empirically interesting phenomena--i.e., that describe phenomena that yield valid generalizations. Sometimes empirical discoveries lead to revisions in such definitions. We now know, for example, that the definition of classical conditioning in terms of contiguity was wrong, and we have replaced it with a definition in terms of information (Rescorla, 1967).

CHAPTER FIVE

FOLK PSYCHOLOGY'S CRITIQUE

The purpose of this chapter is to present an important argument against the cogency of behavior analysis--indeed, against behavioral psychology in general. Among philosophers, at least, this argument has had great influence. If there is a consensus among philosophers about the inherent limitations of the behavioral approach to psychology, this argument is largely responsible for it. Historically, this argument emerged as a response to an argument on behalf of behavioral psychology. Thus, we shall introduce the argument against behavioral psychology by first examining an argument for it.

The Methodological Argument for Behavioral Psychology. The argument that initiates this dialectic purports to show that inner states and events are irrelevant to causal accounts of behavior. This argument can be found in Skinner (1953a), and therefore is sometimes discussed under the heading of radical behaviorism. But its importance is precisely that it does not depend upon radical behaviorism's controversial thesis that mental entities do not exist. It is a methodological argument, and far from assuming that mental entities do not exist, it assumes they do.

It purports to show that no matter what the proximate inner causes of behavior may be (whether they are physiological or mental or both), there is always a valid environmental account of behavior, and this environmental account is superior to the inner account. Paraphrased, Skinner's (1953a, pp. 34-35) argument is this.

> Consider behavior that adjusts to the individual organism's changing environment. If we want to predict and control such behavior, then inner events (whether mental or physiological) are in the final analysis irrelevant. For though the proximate cause of the behavior is presumably an inner event, scientific method requires that we assume that this proximate cause is itself the effect of some cause, which in turn is the effect of some other cause, and so on. If we continue to trace the causal chain backward we eventually arrive at an environmental cause. And such a cause affords a major scientific advantage, for unlike the others, it is located where we can observe, measure, manipulate, and control it. Thus, for any response that adjusts to the changing environment, there is always a causal chain consisting of at least three links: (1) an environmental condition, (2) an inner state, and (3) the behavior. Unless there is a break in the causal chain so that the second link is not lawfully determined by the first, or the third by the second, then the first and third links must be lawfully related. And since the study of this causal connection will maximize scientific progress towards prediction and control, an analysis that connects the first link with the third is to be preferred. But then we may examine the third link as a function of the first, ignoring inner events and looking for the environment-to-behavior causal regularities that scientific method assures us exist.

In summary, "the objection to inner states is not that they do not exist, but that they are not relevant" in a causal analysis (p. 35)--an argument which, if sound, would provide a compelling rationale for the behavioral approach.

The paraphrase makes explicit what in the original is implicit: the argument applies only to behavior that in some sense adjusts to the organism's changing environment. (Skinner's example is an animal's consumption of water.) Any behavior that does not fit this description--that just appears without any relationship to the circumstances of the organism--will slip the grasp of this argument. Hiccups, for example, do not seem to be adjustments to the environment, so this argument would not purport to show that there are environment-to-hiccup regularities. There may turn out to be such regularities anyway, but the argument in question does not insure there will be. What it does attempt to show is that any behavior that would normally fall within the purview of psychology--roughly speaking, behavior that maintains some kind of appropriateness with respect to the organism's changing environment--will follow environment-to-behavior causal regularities.

The strength of the argument is that it concedes from the outset that the entities studied by physiology and mentalistic psychology may in fact be the proximate causes of behavior. Thus, it grants causal efficacy to physiological causes, such as the neural state that accompanies thirst, and to psychic causes, such as the feeling of thirst. But it purports to show that an explanation based exclusively upon environment-to-behavior regularities is nevertheless both possible and preferable, by using its opponent's own premises against him. If there are inner causes, they must ultimately trace back to outer causes. And these outer causes are easier to study and control. Therefore, we might as well take the direct route to a causal account (Skinner, 1967).

Skinner implies that this conclusion is a simple consequence of determinism, but his example makes clear that there is more at work than mere determinism. Determinism would perhaps insure the existence of an outer cause, but without further assumptions it remains possible that this outer cause would be located somewhere the behavioral psychologist would not be searching. The animal has not had a drink for several hours, and therefore is likely to take advantage of an opportunity to get one. The animal has just consumed some heavily salted food, and therefore is likely to take a drink if given an opportunity to do so. These examples of environmental causes are closely tied to the very factors that make the behavior an appropriate adjustment to the changing environment. The organism has causal contact with the aspect of the environment to which it is adjusting. What else would explain the appropriateness of its behavior? So the causal route that passes through the organism's inner states must find its way to the changing environmental circumstances to which the animal's behavior is in some sense an adjustment.

Of course, sometimes an organism's behavior appears to be responsive to changes in one aspect of the environment, but in fact is in contact with some other aspect which co-varies with the first. The hoof beats of Clever Hans seemed to be under the control of the additive properties of the numbers spoken to him, but were later shown to be under the control of his owner's posture and tone of voice. The sequential properties of Nim Chimsky's gestures appeared to be under the control of the syntactic principles to which his linguistic training had exposed him, but upon closer examination it was just as reasonable to assume these sequential properties to be under the control of his trainers' unintentional cues (Terrace, 1979). In the first case, analysis revealed that we had misidentified the part of the environment that was controlling the behavior, in the second it revealed that we could not be sure that our initial identification of the environmental cause was correct. But in both cases we still end up attributing the behavior to an environmental cause. That there is such a cause in all cases in which behavior adjusts to the environment seems relatively uncontroversial. It is difficult to think of a plausible alternative: Uncaused causes? Pre-established harmony? These are not serious competitors with the straightforward thesis that the organism is adjusting to the changing environment as a result of some kind of causal contact with it.

The point of the argument, however, is not merely to show there is always an environmental cause of behavior, but to show there is always an environment-to-behavior causal regularity. The existence of such a regularity is, in fact, the crucial issue. If one always exists, then the cogency of behavioral psychology is guaranteed. For behavior-that-adjusts-to-its-environment will always have an environmental explanation by way of subsumption under such a causal regularity. We may not know what the regularity is, but if the argument is sound, we can be confident there is one waiting to be discovered right where behavioral psychology is searching.

Unfortunately, the argument is not sound. It does not prove there must be environment-to-behavior regularities for all behavior-that-adjusts-to-its-environment. Lacey (1979) and Wessells (1981) have independently arrived at the same refutation, which we may paraphrase as follows.

> Even if it is true that for any given token of a certain type of behavior there must be a chain of causes that goes back through one or more inner states to the environment, this causal chain is a sequence of event tokens. And although it is presumably true that if token *a* caused token *b*, then these event tokens are somehow related to one another lawfully, it does not follow that any event types of which they are tokens are related to one another lawfully. For all we know, each environment-to-behavior causal chain may be due to a unique set of inner circumstances, so that if there were

> even minor changes in these circumstances, an environmental event of the same type as *a* would fail to cause a response of the same type as *b*. Thus even if every response has an environmental cause, that does not mean there are valid environment-to-behavior causal regularities.

The basic point is by no means limited to psychology. Even if a certain event has a certain distal cause, there is no guarantee that there is a regularity that connects the distal cause to its effect. Often the only way to state a valid causal principle is to explicitly take into account the intervening process. But then we cannot necessarily examine the third link as a function of the first, for the third link may be a function of the first only when the second link is in place. Instead of a distal-cause-to-distal-effect regularity, we may get only a regularity that says a certain distal cause will produce a certain distal effect if the intervening process has certain properties. And this possibility cannot give much comfort to the behavioral psychologist.

Consider a non-psychological example of the problem. Suppose a forest fire wipes out the last viable habitat of the spotted owl. The specific event of this forest fire thus causes the spotted owl to go extinct because it sets in motion a chain of events that prevents the surviving population of owls from reproducing successfully. The causal chain connecting the fire with extinction is complex. Some pairs never build nests because they do not find a suitable site, others build nests but are forced to do so in locations that expose their eggs and nestlings to heavy predation, and still others locate their nests in areas that present the adult birds with such fierce competition for prey that they are unable to provide their offspring with adequate nourishment. The result in ten years is extinction. Thus the event of the fire leads, through a complex intervening process, to the effect of extinction.

Does this mean there is a causal regularity connecting forest fires to extinction of species, without any mention of the intervening process? Not at all. Only under special circumstances will fires lead to extinction. But if we include the circumstances as part of the regularity--e.g., the circumstance that the fire will force mating pairs to choose nest sites that expose their offspring to heavy predation--then we find ourselves referring to some of the intervening causes that we supposedly were circumventing.

Now consider a behavioral example. Suppose on a given occasion the sound of buzzing causes Susan to think of bees, and this causes her to think of honey, which causes her to remember she is out of honey, which causes her to put honey on her shopping list. So hearing the buzzing sound causes her to put honey on her shopping list. Does this mean there is an environment-to-behavior regularity which this causal connection exemplifies? Not necessarily. For under different circumstances, buzzing might not have reminded Susan of bees, or bees might not have reminded her of honey, or the thought of honey might not have reminded her that she was out of it, or remembering she was out of honey might not have caused her to put honey on her shopping list. And how are we to distinguish between those other circumstances and the current ones without mentioning inner states or processes? Our search for lawful regularities between environment and behavior seems to lead directly to the intervening causes.

The behavioral psychologist can, of course, counter that these intervening causes are themselves caused environmentally, so there would in principle be a way to locate the relevant circumstances in the environment, rather than inside the organism. But the same problem would be raised all over again, and this time the environmental cause will be harder to locate, because it may have occurred long ago at some other place. The burden of proof has now shifted. Skinner's argument was supposed to show that an environmental regularity always exists. The refutation shows that there is no guarantee it exists, and some reason to think we would have great difficulty finding it even if it did.

So at most, Skinner's argument shows only that behavior that adjusts to changing environmental circumstances has an environmental cause. But it does not show that there is always an environment-to-behavior regularity that connects the environmental cause to the behavioral effect. And it does not show that searching for such a regularity (on the assumption it exists) will be a more direct route to an account of behavior than studying the processes that connect the environmental cause with its behavioral effect. And since a minimal goal of science is to discover regularities (as opposed to merely giving causal accounts of particular events), the argument fails to guarantee the cogency of the behavioral approach.

Folk Psychology's Counteroffensive. Things are worse than that however. For not only has Skinner failed to show there must be environment-to-behavior regularities, but there is reason to suspect there could not be such regularities--not at least for a considerable range of intentional human behavior. This suspicion is based upon considerations similar to those found in the various refutations of logical (or analytical or philosophical) behaviorism first formulated in the late 1950's and early 1960's.

Logical behaviorism is the doctrine that the meaning or content of the mentalistic terms used in ordinary explanations of human action can be analyzed as behavioral dispositions to perform certain physically described responses in certain physically described environmental circumstances. Coveting an object, for example, might be defined as the disposition to handle the object at every opportunity; the belief that an object is booby-trapped might be defined as a disposition to avoid picking the object up or moving it; and so on. The problem with this doctrine is now well known. The disposition accompanying a given desire or belief can be modified by the agent's other desires and beliefs. For example, a person who covets the diamond tiara over there may nonetheless not be disposed to handle it if she believes it is booby-trapped. Even this, however, cannot be said unconditionally. For the state of coveting the tiara and the belief that it is booby-trapped might lead her to walk over and pick the tiara up--if, for example, she was feeling guilty and she wanted to inflict pain upon herself as a form of punishment.

Notice that such changes of disposition are not always a case of one disposition overwhelming another, but of the disposition itself changing as a result of the way mental states interact with one another. Folk psychology provides an endless supply of such examples, in which the introduction of a certain belief or desire reverses or transforms the disposition that had accompanied another belief or desire. Thus, to specify correctly the behavioral disposition that will accompany a given belief or desire, one needs to know the subject's other beliefs and desires. And if one tries to give these other beliefs and desires a purely behavioral definition, the same thing happens. And so it goes.

One never specifies the correct behavioral disposition of a mental state without reference to a certain background of beliefs and desires. So purely behavioral definitions are always incorrect, and correct definitions are never purely behavioral. The hope of logical behaviorists was that we could (in principle, at least) translate mentalistic terms into a non-mentalistic idiom. Apparently, this cannot be done. Thus, logical behaviorism is false. Of course, the refutation of logical behaviorism does not address the question of whether psychology should simply avoid mentalistic terms. But the same premises that refute logical behaviorism can seemingly be used to refute the cogency of any purely behavioral approach to intentional behavior.

Folk psychology sees intentional behavior as the result of the organism's interpretation of its environment in combination with its background beliefs and desires. Why did Ms. Brown drink the poison? Because she believed the glass contained only Chablis, and she loved Chablis. Why did Mr. Jones press the button marked five? Because he wanted the elevator to take him to the fifth floor, and he believed that by pressing the button marked five he would cause it to do so. These are standard folk psychological explanations of human action. They are so much a part of our ordinary understanding of ourselves that it is difficult to conceive human action in any other terms. But the conception of behavior upon which these explanations draw would seem to indicate that any purely environmental explanation of intentional behavior will necessarily overlook important causal factors.

For folk psychology seems to imply that valid environment-to-behavior regularities (at least in the domain of intentional behavior) are dependent upon the context of beliefs and desires. Change the beliefs and desires, and you change the action. If Ms. Jones had believed the glass contained poison, the environmental stimulus of having access to the glass would not have led her to drink the contents. Instead, the same stimulus would have led to the opposite response of avoidance--unless, of course, she had wanted to die, in which case the effect of the stimulus would again have been reversed and would have led her to drink the wine. The initiating cause of Ms. Jones's action in each case could be located in the physical environment, but any attempt to explain the effect of the environment upon her behavior would have to mention the underlying states of belief and desire, for they are what determine the effect.

One finds this argument offered again and again as the central reason why behavioral psychology is incapable of shedding light on human action and why the principles of behavioral psychology are valid only within very limited circumstances--namely, only within circumstances that somehow guarantee the subject will hold certain beliefs and desires. Charles Taylor's influential book, *The Explanation of Behavior* (1964), applies this criticism even to behavioral theories of animal learning. (In the following passage, what he calls the cognitive view is now customarily referred to as folk psychology.)

> On the cognitive view learning is not a function simply of the responses occurring concurrently with the stimuli, or with the stimuli together with reward, but also of the way the stimulus situation is seen by the animal. Thus learning depends not just on the sequence of stimuli and responses but also on the "hypotheses" or "expectancies" that the animal is testing on these trials, in other words on his intentional environment at the time. But then it is possible, and in some situations likely, that the law-like relation posited by S-R theory between response history and current responses will not hold. Thus, to take an example from the range we shall be discussing, that of discrimination learning: The fact that a rat's jumping to what is in fact a white card is followed by reward may not serve to strengthen the tendency to jump to the white card in [the] future. For the rat may not have been "paying attention" to the color of the card but might have been "testing the hypothesis" that jumping to the right-hand card brings reward. In other words, the "intentional description" under which the rat jumped to it was "card on the right-hand side" and not "white card". But then if the "solution" is jumping to the white card, i.e. if reward varies only randomly with position, but is constantly linked with color, this trial will not have helped in any way to strengthen the correct response, even though the card jumped to was white and reward followed.
>
> Hence on the cognitive view it matters what the rat is *doing*, that is, what action he is performing, and thus what intentional description the action has for him, whether "jumping right" or "jumping to white", whereas on the S-R view, the response is not an action, the intentional description is irrelevant, and it matters only what descriptions the card actually bears to which the rat jumped. (pp. 124-125, italics in original)

Taylor is saying that if we are to give a maximally complete account of the animal's behavior, our explanations must refer to the animal's interpretation of its environment. Otherwise we shall in effect have a set of principles which apply only to situations in which the animal interprets its environment in a certain way. But if we are to arrive at principles that handle all possible circumstances, we need to acknowledge the animal's interpretation of its environment as one of the factors we take into account. But such an account will no longer be behavioral. Q.E.D.

Daniel Dennett makes a similar point in *Brainstorms* (1978).

> One can be a behaviorist in explaining and controlling the behavior of laboratory animals only so long as he can rely on there being no serious dislocation between the actual environment of the experiment and the environment perceived by the animals. A tactic for embarrassing behaviorists in the laboratory is to set up experiments that deceive the subjects: if the deception succeeds, their behavior is predictable from their false *beliefs* about the environment, not from the actual environment. (p. 275, italics in original)

Like Taylor, he suggests not so much that behavioral principles are invalid, but that they are valid only up to a point, and furthermore, that this point is defined cognitively--i.e., by non-behavioral factors. Thus, behavioral psychology cannot define the (very limited) circumstances under which its principles are valid without ceasing to be behavioral.

Jerry Fodor carries this discussion a step further by pointing out that the crucial thing about beliefs and desires is the way they interact.

> Mental causes typically give rise to behavioral effects by virtue of their interaction with other mental causes. For example, having a headache causes a disposition to take aspirin only if one also has the desire to get rid of the headache, the belief that aspirin exists, the belief that taking aspirin reduces headaches and so on. Since mental states interact in generating behavior, it will be necessary to find a construal of psychological explanations that posits mental processes: causal sequences of mental events. (1981, p. 289)

An immediate corollary is that virtually any physical response is compatible with virtually any physical stimulus, depending upon the organism's beliefs and desires (Fodor, 1975, p. 63). So, reliable prediction and control seems intrinsically beyond behavioral psychology's grasp. If its principles turn out by and large to conform to actual behavior, this is just a lucky coincidence.

One might wonder if this problem could be circumvented by means of environment-to-mental-state correlations. Could not behavioral psychology take mental states into account by referring to their environmental causes? If a certain belief arises from a certain stimulus, then instead of mentioning the belief, mention the stimulus.

Unfortunately, this strategy simply moves the difficulty back one step. If we are to replace reference to mental states with reference to their environmental causes, there must be valid regularities which say that such-and-such environmental event always leads to such-and-such mental state. But if folk psychology is correct, such regularities are valid only if the context of beliefs and desires is just right. Thus, we run into the same problem--we cannot define valid regularities without explicitly mentioning the beliefs and desires of the organism. No matter what environment-to-mental-state regularity we might consider, there will be indefinitely many belief/desire combinations that would disrupt the regularity.

Perhaps one could defend the validity of behavioral principles by adding ceteris paribus clauses. Few sciences have causal principles saying that whenever such-and-such occurs it is followed by so-and-so. Newtonian theory, for example, does not say that a force of such-and-such size will move an object of such-and-such mass, but just that it has a tendency to do so. It explicitly leaves open the possibility that this tendency will be counteracted or even overwhelmed by an opposing force of equal or greater intensity. Behavioral principles, it might be argued, should be understood in a similar manner. A hungry animal has a tendency to eat available food, but its thirst may overwhelm its hunger and cause it to drink rather than eat. More generally, behavior is the result of multiple causation. Thirst will produce a tendency to drink liquids, but this tendency can be countered by the belief that the liquid in this glass is laced with poison. The outcome may be a failure to drink the liquid, but this does not contradict the principles of behavioral psychology--not, at least, when they are correctly understood. Or so it could be (and has been) argued.

But the critique put forward by folk psychology is stronger than this defense can parry. It is true that a hungry animal may fail to eat available food because its thirst overwhelms its hunger. And in general, examples in which one desire can overwhelm another are analogous to the way one physical force can overwhelm the effect of another. But the way beliefs and desires can interact to disrupt environment-to-behavior regularities does not fit the countervailing force paradigm. A force imparts a certain tendency to the system upon which it is acting, and this tendency continues to exist even if its effects are counteracted by an opposing force. A push to the south can be nullified by an equally powerful push to the north, but the southward push nonetheless had a southward effect: if there had not been a southward push, the system would have gone north. A push to the south can also be modified, as when it interacts with a push to the east, producing a southeastward movement. Again, the southward push had a southward effect: if it had not occurred, the system would have moved straight east. The point is that the tendency imparted by a force is stable. A southward push does not on some occasions impart a tendency to move eastward. It always imparts a tendency to move southward, although other forces can modify the net result of this tendency.

But folk psychology seems to imply that a given environmental factor is capable of imparting any tendency whatever. This may not be true of every environmental cause. A shrill sound may be inherently aversive, and may impart a tendency for the muscles to stiffen no matter what the animal believes and desires (although repetition of the sound will diminish this tendency). But this is not the type of environmental effect that is central to the explanation of intentional behavior. What tendency does the sight of a glass of wine impart? There does not seem to be an answer that is independent of the beliefs and desires of the subject. But if environmental factors take on and lose tendencies in ways that are under the control of the organism's system of beliefs and desires, then there would seem to be an inherent incompleteness to the behavioral approach. Its generalizations will be valid only to the extent that the organism to which they are applied has a certain belief/desire profile.

Common sense seems to acknowledge this in the way it attributes general tendencies to human subjects. For example, common sense expects people who have a prominently displayed gun pointed at them to fear for their lives. Why? Because most people would have certain relevant beliefs and desires. They would believe the gun to be pointed at them; they would believe it might be loaded; they would believe that with the slightest tension on the trigger, the gun could kill them; and they would desire not to be killed. But common sense also hedges its prediction with certain qualifications. Other beliefs and desires are possible, and these would lead to a different response. Someone might interpret the gun to be a comic prop from which an American flag will appear, and therefore be amused instead of fearful. Someone else might think the gun is real but believe it not to be loaded, and therefore feel contempt rather than fear. And yet another person may believe the gun to be real and loaded, but want to die, and therefore experience hope.

As a result, common sense hedges every generalization about the effects of a given type of environmental factor with qualifications that basically say: an environmental event will induce a tendency to perform a certain act, in an agent having a normal set of beliefs and desires. The term *normal* carries a lot of weight, and its meaning changes from culture to culture, and even from context to context. Pointing a loaded gun at someone does not in itself generate a certain tendency. Such a stimulus may induce any imaginable tendency, given the right set of beliefs and desires. For everyday purposes, we often can get a handle on this set of beliefs and desires by knowing the background of the person and the nature of the immediate situation. But even then, we must assume the person to be normal and to be making a reasonable interpretation of the situation.

There seem to be no intrinsic tendencies of action on the part of an organism that has an unbounded set of potential beliefs and desires. In such an organism, a change of belief or desire can cancel, reverse, or modify almost any environment-to-behavior tendency in any number of ways. This suggests that the behavioral approach will arrive at valid behavioral generalizations only to the extent that the beliefs and desires relevant to behavior are heavily constrained. Essentially, the critique of behaviorism offered by Taylor, Dennett, Fodor and others maintains that if folk psychology is true, then the validity of environment-to-action regularities is limited by the extent to which underlying beliefs and desires are constrained.[1] This implies that behavioral regularities are valid only to the extent that these underlying factors do not conspire against them.

So long as one studies a rat or a pigeon, and constrains what ends it can pursue, and what means it may employ to gain these ends, and provides it with obvious and reliable signals about what means are currently effective, then the organism will behave in the ways behavioral psychology predicts. But if one applies behavioral principles to human beings, then they are less apt to be valid. They may work reasonably well for infants and young children, and perhaps even for developmentally disabled adults, but only because the underlying systems of beliefs and desires are so highly constrained. These principles may also be valid within closed environments such as factories, prisons, public schools, and bureaucracies, but this is only because these environments mimic the highly constrained conditions of the conditioning chamber, where there are limited goals to pursue, limited means for pursuing them, and a few obvious signals about what means are currently effective (Schwartz & Lacey, 1982). But put a normal adult human being in a complex environment, where there are no contrived limitations on the number of goals that can be pursued, no artificial constraints upon the variety of means that can be used to pursue them, and no obvious, foolproof signals about which of these means is optimal, and none of the alleged principles of behavior will hold true, because they do not take into account the complex way an unbounded system of beliefs and desires interacts with environmental stimuli. Whatever the supposed environment-to-behavior regularity

---

[1]A moment's reflection reveals that this qualification applies as much to mental factors as to environmental ones. There is no specific action that a certain belief or desire tends to produce. For the tendency produced by a belief or desire is always the result of its interaction with background beliefs and desires. What tendency, for example, is produced by belief in an afterlife? It depends upon what else the subject believes and desires.

may be, there are indefinitely many belief/desire combinations that will disrupt it. And the less constrained these combinations are, the more likely there is to be one that disrupts the behavioral regularity. So the more a person's higher cognitive capacities develop and the richer a person's environment becomes, the less relevant behavioral principles are to understanding that person's intentional behavior.

This argument is often combined with the claim that the basic principles of behavioral psychology are tautologies. The result is a dilemma. Either the principles of behavioral psychology are tautologies or they make empirical claims about behavior. If they are tautologies, then they are of no use for purposes of prediction and control. If they make empirical claims about behavior, then the critique of folk psychology shows them to be of only limited validity, and once again they are of no use (or very severely limited use) for purposes of prediction and control of normal adults in complex environments. Therefore, they are of very limited use for purposes of prediction and control. And since prediction and control is supposed to be behavioral psychology's strong suit, this puts the entire field on the defensive.

CHAPTER SIX

REBUTTING FOLK PSYCHOLOGY'S CRITIQUE

We have examined a family of arguments purporting to show that behavioral methods are incapable of giving a valid account of the intentional behavior of normal human adults in the complex settings of everyday life. These arguments may be summarized as follows.

> Folk psychology tells us that a subject's beliefs and desires explain his or her intentional behavior. To the extent that such behavior is subject to environment-to-behavior regularities, these regularities are conditional upon mental factors. Therefore, unless we limit behavioral principles to simple organisms or to highly constrained environments (and thereby contrive to hold the underlying factors constant), the only valid principles relating the environment to behavior would have the form: if the subject is in such-and-such mental state, then so-and-so environmental condition leads to thus-and-so response. But this is no longer a behavioral principle. Thus, there are no valid context free principles of intentional behavior that do not refer to underlying mental factors.

This generic argument, used by Taylor, Dennett, Fodor, Schwartz & Lacey, and others, borrows premises from folk psychology to conclude that unless there is a stable background of beliefs and desires, the environment is not lawfully related to intentional action (or to operant behavior, which for purposes of this argument is taken as equivalent to intentional action).

The argument itself makes clear the sense in which behavioral methods are held to be flawed. It is not that there are questions about intentional behavior which are beyond the scope of behavioral methods (in itself, not necessarily a decisive objection to a scientific program), nor that non-behavioral forces can intrude upon, and overwhelm, behavioral forces (behavior analysts would concede this can happen, but ask how this is different from the situation in many other sciences?). It is instead that there are factors intrinsically beyond the scope of behavioral methods that control the very regularities behavioral psychology studies. Change certain of these factors, and behavioral regularities change drastically, sometimes even to the point of reversing themselves so that a factor that was increasing the value of a dependent variable now decreases it. As a result, once behavioral psychologists venture beyond simple organisms and simple environments, they never know whether an environmental factor will add to, subtract from, or be neutral with respect to, the probability of some response--unless, of course, they (perhaps surreptitiously) hold constant the underlying mental factors that behavioral theories attempt systematically to finesse.[1] The purpose of this chapter is to evaluate this family of arguments. Let us begin with some that fail.

I

One strategy for responding to this argument is to assert that the mental states upon which the behavioral regularity is conditional must themselves have environmental causes. If the behavioral regularity says that environmental factor E leads to response R, and this regularity is conditional upon some mental state M, then the valid, context-free regularity will be of the form: If M, then E leads to R. But (the argument goes) there must be a law that connects some environmental cause, C, to M. This law will have the form: if C then M. By the logical principle of transitivity, we can now replace the mentalistic antecedent with an environmental one, giving us a

[1]These arguments do not attack the legitimacy of behavior analysis so much as attempt to contain its significance within certain boundaries. They readily concede that behavioral methods are valuable when applied to children, to the developmentally disabled, to victims of autism, and (more controversially) perhaps even to people who are behaviorally deviant. But what the preceding arguments attempt to show is that the closer we get to normal adult human behavior, the less adequate behavioral methods become. And when we reach highly skilled, creative behavior, their utility vanishes altogether.

behavioral principle that says: if C, then E leads to R.[2]

The problem with this argument is that there is little (if any) reason to assume there will be such an environment-to-mental-state regularity. The same problems that arise for environment-to-behavior regularities arise for environment-to-mental-state regularities--i.e., unless we hold the context of mental states constant, we cannot expect these regularities to remain valid. Of course, in the case of simple organisms within simple environments, the mental context may remain constant just by dint of the simple-mindedness of the organism and the aridity of the environment. But if we consider normal human behavior in a rich social environment, the context of underlying mental states can vary widely enough to disrupt virtually any environment-to-mental-state regularity. But to take these mental factors into account would require us to cease following the behavioral method. And it is no use arguing that these mental factors are also under environmental control, because we would only face the same problem at a third level, and so on. Evidently, this is not a promising line of defense.

A more direct and aggressive defense would be to launch an attack against the principles of folk psychology upon which the critique is based. These principles may seem a rather flimsy basis upon which to build a critique of a progressive scientific program. After all, the history of science is an unbroken series of defeats for the folk theories of physics, chemistry, biology, etc., so why expect folk psychology to be any different? Indeed, the defeat of the folk theory in a given domain is often the crucial breakthrough leading to accelerated scientific progress. So one might suspect folk psychology of being an obstacle to scientific progress, rather than a corrective for scientific error. True, as some philosophers (e.g., Dennett) have emphasized, folk psychology is less easily set aside than other folk theories. It may in some sense be irreplaceable for certain purposes such as ethics, law, and perhaps even meta-science. But none of this refutes the expectation that it will ultimately be abandoned as a basis for scientific accounts of behavior.

On the other hand, even though the demise of folk psychology (qua

---

[2]One finds this argument in Lacey & Rachlin (1978), Skinner (1953a), and many other locations.

science) may be inevitable, the question nonetheless arises as to whether its successor will retain those of its properties which raise difficulties for behavioral principles. Although this is a question no one can answer with authority, I think we can make a plausible guess.

Let us begin with a reasonable surmise: Indispensable as they may be as multipurpose tools for everyday purposes, folk psychological concepts individuate mental states in a way that does not capture lawful regularities with the precision expected of science. Consider its system of character traits. We are not surprised to find a shy child who nonetheless is bold in certain social settings, a courageous soldier who is afraid to speak in public, or a brilliant scientist who cannot decipher train schedules. Such examples seem to indicate that folk psychology fails Aristotle's test of carving nature at its joints. Thus, despite the availability of an extensive vocabulary of character traits, we can often create a more accurate set of expectations about a person's behavior by telling a few stories about him than by presenting a long list of his character traits.

Now consider the folk psychological concept of an attitude towards a proposition. The millenarian believes that the second coming of Christ is about to occur, and so he gives his worldly goods to the poor and repairs to the hilltop to witness the final days. The patriot desires that her nation accomplish great deeds, and so she attends the rally to inspire the troops. Such states of mind seem to fare no better than shyness, courage or brilliance. The category of belief, for example, does not distinguish among beliefs based upon direct experience, inference, and observation of others. Each of these affects behavior in a different way and therefore needs to be assigned to a separate category.

This point is closely related to one of the central themes of modern psychology: There are subsystems of the mind, the content-bearing states of which do not freely interact with those of other subsystems. Folk psychology knows nothing of these, although scientific speculation about the mind, from Freud, who posited an ego, superego, and id, to Fodor, who posits a language faculty, a face-recognition faculty, a geometrical faculty, etc., has made the delineation of subsystems the centerpiece of theoretical psychology. For these and other reasons, philosophers have speculated that there is something deeply flawed in the way folk psychology describes the mind (cf., Stich, 1983).

But just how troublesome is this for the preceding critique of behavioral psychology? We may have character traits, even if not the ones everyday vocabulary assigns us. And we also may have states that interact on the basis of their propositional content, even if those states are not the ones folk psychology attributes to us. It is the logic of the way content-bearing states interact, not the way we individuate them, that causes problems for behavioral psychology. So long as psychology is committed to states that interact on the basis of their content (or on the basis of some property that is roughly isomorphic to content), then the key assumption for the critique of behavioral principles will remain intact. And it is precisely this assumption that inner states have content, and interact in ways that reflect this content, that contemporary cognitive science shares with folk psychology. Thus, there may be little comfort for behavior analysts in the forthcoming demise of folk psychology. Whatever replaces folk psychology is likely to share the features of its predecessor that have caused the behavioral approach so much grief. So for our purposes, we might as well assume folk psychology to be true. The aspect of it that is relevant to the critique probably is.[3]

II

This however does not necessarily mean that we must accept the critique's conclusion, for even if the premises are true, they may not support the conclusion. Indeed, as we shall attempt to show, the argument is a complex non sequitur. It may refute the behavioral approach to psychology as Hull and others once conceived it, or as

[3]It may be worth pausing to note that none of the conclusions of this essay depend upon the truth or validity of folk psychology. The relevance of folk psychology to this analysis is to serve as an obstacle to the attempt to find a legitimate place for operant conditioning in contemporary psychology. Many philosophers believe that if folk psychology is even approximately true, it poses a decisive challenge to the relevance of operant conditioning to normal human behavior.

logical behaviorists once imagined it, but not as Skinner conceived it, and as behavior analysts practice it. Critics of behavior analysis have misunderstood what operant principles are saying. Once we remove these misunderstandings, little is left of the critique.

The Dependent Variable Extends Over Time. Let us begin by recalling that the dependent variable of a behavior analysis is not the occurrence of a response, but is some aspect of responding (usually rate) over a relatively extended period of time (Malone, 1987a; Hinson, 1987). The principle of the fixed-interval scallop, for example, says that if an organism comes under the control of a fixed-interval schedule long enough to reach a steady state, then the pattern of responding will take a certain form. What this explains is not the individual response, but the changing rate at which responses are performed. Folk psychology does not normally address this question. For example, it gives little (if any) guidance on the question of what stable pattern of responding will eventually emerge on a fixed-interval schedule. Once that pattern has been discovered, of course, folk theory may offer an explanation of what beliefs and desires must underlie this pattern (indeed, it seems capable of offering a plausible explanation of virtually any pattern of behavior--which is one reason why behaviorists have treated it with suspicion), but this is quite different from contradicting or refuting or anticipating the principle.

The Independent Variable Extends Over Time. Given that critics are sometimes unaware that the dependent variable extends over time, one is not surprised to find they are sometimes also unaware that the independent variable does so too. This is important, because the tendency of folk psychology is to look at a particular response to a given stimulus and note that if the organism were to change a certain belief, then the same stimulus would result in a different response. While this is no doubt true, to take this as a refutation of behavioral principles is to overlook the temporal dimension of the independent variable. If certain beliefs would interfere with the ability of the contingencies of reinforcement to move the organism toward the usual steady state, then presumably they would tend not to arise, or if they do arise, their effect upon behavior would tend to be neutralized by other adjustments within the organism. Behavior analysis does not say how these adjustments occur. It does not even say whether one or many mechanisms are responsible for the behavioral patterns captured by its principles. It simply specifies what pattern will eventually emerge if the organism is exposed to the contingencies for a considerable duration of time.

The relevance of this temporal dimension of the independent variable is particularly striking in a version of the critique of behavioral psychology due to Jerry Fodor. Fodor (1981) asks us to imagine two coke machines that charge ten cents per bottle. The behaviorist machine takes dimes only. So when you put a dime in, a coke comes out. This is Fodor's model of a behavioral regularity. We do not need to know what goes on inside the machine to know that a dime input causes a coke output. Fodor has the behaviorist concluding that inner states are irrelevant to a causal account of coke machines.

But unfortunately for the behaviorist, there is also a mentalist coke machine, and this one can take dimes or nickels. If you put a dime in, a coke comes out. But if you put a nickel in, one of two things can happen: if it is the first nickel deposited since a coke came out, the machine will enter a state of readiness; if it is the second nickel deposited, the machine will dispense a coke. Fodor calls this the mentalistic coke machine, because he thinks it illustrates at a very elementary level the usefulness of inner states.

Before criticizing this argument, let us note the sense in which it does succeed in making contact with the question under discussion. Fodor is not simply claiming that inner states are necessary to give an account of the process underlying behavioral regularities. He is claiming that the regularities themselves cannot be valid unless they make reference to inner states. A behavioral theory of coke machines cannot give an accurate causal account of the mentalistic machine's behavior, because it cannot distinguish between situations in which a nickel input will produce a coke and those in which it will not. Only by positing an inner state of readiness or something of the sort, can we accurately describe the regularity regarding nickel inputs. When the machine is in the state of readiness, a nickel produces a coke; when the machine is not in the state of readiness, a nickel causes it to enter the state of readiness (but does not produce a coke). By this example, Fodor seeks to illustrate the basic limitation of behavioral principles.

What he fails to notice, however, is the isomorphism between an inner state such as readiness and a certain environmental history. The state of readiness is nomically equivalent to an environmental history of a nickel having been deposited but not having been followed by delivery of a coke. Because of this equivalence, the behavior analyst can describe the mentalistic coke machine as follows: when a nickel is deposited after a coke has been dispensed, the machine does not produce a coke; when a second nickel is deposited, however, the machine does produce a coke.[4]

---

[4]R.J. Nelson (1969, 1982, 1984, 1989) has suggested that this type of isomorphism is the basis for neobehaviorism, which acknowledges the need for mental states but claims they are nomically equivalent to environmental histories. I am sympathetic to Nelson's desire to show there is a certain kind of compatibility between mental and behavioral theories. His own treatment is of such complexity and subtlety that I cannot do it justice here. I am not however inclined to believe that the isomorphism in question extends to all cognitive states of interest to mentalistic psychology--i.e., I suspect that Nelson has overgeneralized.

Consider, for example, Fodor's mentalistic coke machine. We can give a behavioral definition of the readiness of the machine to dispense a coke (the output) upon the deposit of a nickel (the immediate input) by taking into account the history of past deposits (distal inputs). Nelson has shown how to extend this result to all automata, concluding that if the mind can be modeled as an automaton (i.e., as a Turing machine), then there is an isomorphism between mental states and environmental histories.

But it is questionable whether the mind can be modeled as a Turing machine. To see why, return to Fodor's coke machine. As a model of mind, the feature of the machine that makes it so easy to turn into a Turing machine is its resoluteness of desire. If the machine were a living being, it would resemble a creature that wanted only one thing, and always wanted it--namely, to dispense cokes. The problems for the program of producing behavioral equivalents of mental models only get interesting (i.e., overwhelmingly complex) when we confront a mind with complex beliefs and complex desires. What Nelson has done is show that the dispositions of a mind with complex beliefs and one simple desire can be modeled behaviorally, if we permit the antecedents of our dispositional regularities to stretch out across time. He has not, however, shown this to be true for a mind with complex beliefs and desires, and I doubt whether this can in fact be done.

In general (i.e., except for simple organisms in simple environments), behavioral regularities and mental regularities are non-equivalent. The relationship between the

Fodor's counter-example would work only if the independent variables of behavior theory had to be temporally contiguous with the dependent variable. But they are not thus limited (Cf. Lacey & Rachlin, 1978). They stretch out across time. The principle of the fixed-interval scallop tells us, for example, what pattern of behavior is maintained by extended exposure to a fixed-interval schedule. The independent variable is not simply the delivery of the reinforcer in the presence of the discriminative stimulus, or even such delivery in the context of a lapse of a fixed-interval of time since the last delivery of the reinforcer, but is the repeated exposure to this complex stimulus over an extended period of time. Thus, the principle connects a certain pattern of behavior that extends over a sizable duration of time (the scallop) with a cause that also extends over a sizable duration of time (the schedule).[5]

The Descriptive Categories are Not the Variables. We have seen that it is necessary to distinguish between the categories by which behavior is described and the variables that are the causes and effects connected by behavioral principles. The functional categories that a behavior analyst uses to describe the environment/organism system are not themselves the variables in behavioral regularities. An operant response is not itself a dependent variable; a discriminative stimulus or a reinforcing stimulus is not itself an independent variable. Instead, these are the categories that define the behavioral entities whose measurable properties are the variables (cf. Morse & Kelleher, 1977).

A typical operant regularity takes the following form. It says that if response O functions as an operant and if stimulus D functions discriminatively and if stimulus R functions as a reinforcer, and if these entities are related by a three-term contingency of reinforcement, then a certain type of contingency will (once the organism has been

---

two is not semantic. It is explanatory. Mental regularities explain behavioral ones.

[5]One might ask whether the organism's historical experience with the schedule is not in some way equivalent to a mentalistic belief or state. Clearly there is a relationship between the two, but the relationship does not seem to be that of equivalence. (See the preceding note for a discussion of this question.)

exposed to the contingency long enough to settle into a steady state) maintain a certain pattern of operant responding (which pattern will have a characteristic shape when represented on a cumulative record). One such contingency is the fixed-interval pattern of reinforcement, which produces scalloped cumulative records; another is the fixed-ratio pattern, which produces stair step records; etc.

Avoiding Folk Psychology's Counterexamples. The preceding analysis of behavioral regularities has immediate implications for the vexing counterexamples that folk psychology poses for behavior theory. The strategy of these counterexamples is to argue that changes in belief and/or desire will disrupt behavioral regularities in ways that a purely behavioral vocabulary cannot accommodate. Let us concede that this argument is effective against S-R psychology. It is not, however, effective against behavior analysis.

Consider the following example. A pigeon is released into the experimental chamber. As a result of the animal's history, key pecks function as operant responses, illumination of the key functions as a discriminative stimulus, and access to grain functions as a reinforcer. We have programmed the equipment so that when (and only when) the key is red, the pigeon gets three seconds of access to grain for the first key peck after an elapse of fifty seconds since the last period of access--i.e., the pigeon is on an FI-50 schedule. Assume the pigeon's behavior stabilizes into the usual fixed-interval scallop. Now suppose we are able to induce a change in the pigeon's desires (it does not matter how). Instead of desiring to consume grain, it now desires nothing but to escape from its immediate environment. So instead of pecking the key, it turns from the key and raises its wings. Does this refute the principle of the fixed-interval scallop? Not at all. When the pigeon's desires changed, access to grain lost its reinforcing function. As soon as grain ceased to be reinforcing, the antecedent of the principle was no longer satisfied, and therefore the principle ceased to entail anything about what the animal would do. The fact that the pigeon quit pecking the key, therefore, does not contradict the principle.[6]

Similarly, suppose we are able to induce the pigeon to believe that red illumination on the key no longer conveys information about the effectiveness of key pecks, although in fact the equipment continues to be programmed as before. Again, the pigeon's behavior is disrupted. Instead of producing a scalloped pattern on the cumulative record, it produces an irregular line. Does this refute the principle of the fixed-interval scallop? No, because when the pigeon's belief changed, so did the function of red illumination, which ceased at that point to be a discriminative stimulus. Thus, the antecedent of the principle was no longer satisfied, and it ceased to have implications about the behavior of the pigeon.

Evidently, then, the functional categories appearing in behavior analytic principles are made to order to solve exactly the problem folk psychology raises. Induce an organism to change its beliefs or desires in a way that alters its operant behavior and one has simultaneously altered the function of certain aspects of the organism/environment system. In disrupting the animal's behavior one has also changed the functional categories that describe the organism/environment system. One therefore has not disproven the targeted behavioral principle, because the principle is conditional upon the applicability of the functional categories.

Parity for the Functional Concepts of Behavior Analysis. Sometimes critics seem completely unaware that functional concepts play any role at all in behavioral generalizations. This is especially evident in a criticism (made in both Taylor, 1964, and Dennett, 1978) to the effect that behavioral generalizations can be refuted by the case in which an organism misinterprets its environment. This line of criticism completely ignores the role of functional concepts, which insure the organism is maintaining a certain type of contact with the environment. When it is not in such contact, the principles do not apply. This is a straightforward consequence of functional concepts, which apply to

---

[6]The issue under discussion at present is not whether mental states play a legitimate role in psychology (behaviorism) but whether the validity of behavioral principles would be undermined if they do.

some entity only if that entity can control or can be controlled by certain events. This is one of the reasons why philosophers have advocated a functional interpretation of mental concepts. But we cannot say functional concepts have this feature in cognitive psychology and then deny them this feature in behavioral psychology. Mere parity of treatment for the functional concepts of behavior analysis will suffice to avoid folk psychology's counterexamples.

The advantages of functional concepts for cognitive psychology are often illustrated by talking about computers. A computer program is a set of functional relationships. To instantiate this program in a given physical system, it may be necessary to construct a novel physical realization of it: one system uses tapes, another floppy disks; one uses vacuum tubes, another transistors. The end result is that functionally equivalent operations are induced in the physically different systems through physically different means. Nonetheless it is true to say they can run the same program. This is the familiar point (due originally it seems to Hilary Putnam) that functional states are not individuated physicalistically.

There is another feature of functional states, however, which is more relevant to the current topic. If a computer overheats, it will not run the program correctly. But this does not imply that no valid regularities about the computer's operation can be formulated with the functional concepts used to describe computer programs. It only means that such regularities are applicable only on the condition that the computer is running properly--i.e., on the condition that its physical states function in a certain way. It goes without saying that if the computer overheats, then the program ceases to be in control and all bets about output are off. We do not however take this to mean that a computer program cannot really determine a computer's output. The program does determine the output--under normal circumstances. We understand this about computer concepts, and take explanations couched in terms of programs accordingly. What is not so well understood is that behavior analytic concepts are also functional in exactly the same sense.

If an organism has grossly inaccurate beliefs (due perhaps to a misperception of the environment), this is analogous to a computer that is overheated. Take a pigeon for which the redness of the key functions as a discriminative stimulus marking the presence of a fixed-interval schedule. Suppose some event--it does not matter what--causes the pigeon to believe that red is no longer correlated with the relevant contingency of reinforcement. Does the principle of the fixed-interval scallop imply that the pigeon will nonetheless respond to the red key with a scalloped pattern of responding? No, because when the pigeon changes its belief with respect to redness of the key, redness ceases to function as a discriminative stimulus marking a fixed-interval schedule.

Or consider a change in belief about the instrumentality of the key. Suppose the pigeon is responding to a red key in the predicted manner when for some reason it ceases to believe that pecking leads to delivery of food. The pigeon stops pecking the key. But this cessation in pecking does not refute the principle of the fixed-interval scallop, because at the same time as there is a change in belief there would also be a change in the function of key pecks. So as long as the functional concepts of behavior analysis receive the same courtesies as do the (supposedly) functional concepts of cognitive psychology, the counterexamples that derive from folk psychology pose no threat to the validity of behavior analytic regularities.

III

There is a sense, then, in which the critique deriving from folk psychology misses the mark--at least so far as behavior analysis is concerned. Having said this, however, we must concede that there is a disturbingly tight fit between the situations in which operant psychology has in fact successfully been applied to human behavior and the situations in which folk psychology's critique implies it should successfully apply--namely, in highly constrained environments such as prisons, factories, elementary classrooms, etc., or with reference to highly constrained human subjects such as young children or developmentally disabled adults. So even if the potential validity of operant principles is not directly refuted by folk psychology's critique, there remains the troubling possibility that the best explanation of operant psychology's strengths and weaknesses, as applied to human behavior, is that its principles are valid only in these limited contexts (Schwartz & Lacey 1982, 1988). One might call this the inductive version of folk psychology's critique, to distinguish it from the deductive version that we have been discussing up to now. The deductive version purports to refute the validity of operant principles, and once this purported refutation is revealed to be a non sequitur, the argument carries no strength. The inductive version, on the other hand, purports not so much to refute operant principles as to call them into question. And so, if the following defense against this argument is not conclusive, it should be born in mind that the argument itself does not purport to be conclusive either.

A Theory of Forces. Behavior analysis takes the behavioral capacities implicit in the concepts of an operant response, a reinforcer, discriminative stimulus, and a three-term contingency that interrelates these three, as primitive concepts (i.e., as concepts in need of no further analysis by the operant psychologist) and proceeds to ask (i) what are the environmental causes of these capacities (i.e., what leads to their acquisition--which is not the same as asking what process underlies their acquisition), and (ii) how do these capacities interact (which is not the same as asking what process underlies their interaction).

The major successes of operant theory to date have been its answers to (ii). These include Skinner's pioneering work on the effects of different schedules of reinforcement, as well as the more recent work by his students and colleagues in the so-called quantitative analysis of behavior (the study of behavioral contrast, concurrent schedules, animal foraging, delay of reinforcement, the economic analysis of behavior, and so on--about some of which more later). These are all examples of behavioral regularities that generalize surprisingly well from situation to situation, from individual to individual, and sometimes even from species to species.

Of course, only regularities described in a certain way are being claimed to generalize in this way--a point which Skinner makes when he comments upon the virtually identical cumulative records produced by a pigeon, a rat, and a monkey. As he says, "Once you have allowed for differences in the ways in which they make contact with the environment, and in the ways in which they act upon the environment, what remains of their behavior shows astonishingly similar properties" (Skinner, 1956, p. 95). It is unlikely that the operant behavior of pigeons, rats, and monkeys is underlain by the same cognitive mechanisms. Nonetheless, operant theory describes behavior in a manner which captures a regularity that these presumably dissimilar mechanisms (whatever they may be) have in common.

This is not to say that all behavioral principles generalize so well. Few answers to (i) do, for example. There is, however, a reasonable explanation for this failure that is based upon the sense in which operant theory can be interpreted as a theory of forces.

Forces are a special type of causal factor that can supplement or counteract the effects of causal factors of the same type. An impressed force, for example, can supplement another impressed force (as when a wind out of the north supplements a southward shove), or it can counteract another impressed force (as when a wind out of the south counteracts a push to the south). The causal factors referred to by operant concepts enter into analogous relationships. One source of reinforcement, for example, can supplement the effect of another (as when the reinforcing approval bestowed by a parent upon a child's academic accomplishments supplements the inherently reinforcing effect of mastering some task), or it can counteract the effect of another (as when the unintentionally reinforcing parental attention that occurs immediately after a child's uncooperative behavior serves to counteract the effect of the punishment the parents administer). Sober (1984) has noted that a theory of forces contains two types of laws: source laws which describe the circumstances that generate forces, and consequence laws which describe how these forces, once generated, produce change in the systems they impinge upon. Sober claims that in the case of evolutionary theory, the consequence laws connect one supervenient property (e.g., fitness) with another supervenient property (e.g., reproduction), and therefore apply to a wide range of circumstances. The source laws, on the other hand, connect physical circumstances (e.g., a certain morphology) with supervenient properties (e.g., fitness). Since there are indefinitely many physically distinct ways of generating these supervenient properties, there are few broad generalizations among the source laws of evolutionary biology

(Sober, 1984, pp. 48-52).[7]

An analogous interpretation may be made of operant psychology. The consequence laws of operant theory (answers to (ii) above) formulate the effects of the three-term contingency upon behavior, in all its many variations (e.g., fixed-interval schedules, fixed-ratio schedules, variable-interval schedules, variable-ratio schedules, concurrent fixed-interval schedules, concurrent variable-interval schedules, etc.). As we have seen, these principles generalize surprisingly well. The source laws (answers to (i) above), on the other hand, identify the circumstances that lead to the three-term contingency (e.g., that cause a stimulus to function discriminatively, that cause a consequence to become a reinforcer, that cause a sequence

---

[7]Sober (1984, p. 48) defines supervenience as follows:

> A property is said to *supervene* on a set of physical properties if it satisfies two conditions. First, the property must not itself be physical, in that different objects may share the property and yet be physically quite different. Second, two systems that are physically identical must both have or both lack the property in question.

Recent philosophical work calls into question the appropriateness of Sober's attribution of supervenience to evolutionary concepts, but for reasons that do not affect the validity of the point he makes about capturing broad generalizations. (See, for example, Kim, 1984; Teller, 1984; Dretske, 1989; and Enc & Adams, 1992.)

Even if fitness is not a supervenient concept, it applies to indefinitely many physically distinct systems. And it is this feature, not supervenience *per se*, that renders a concept capable of formulating generalizations that purely physical concepts would not capture. The reader who is curious about technical philosophical issues relating to supervenience and psychological explanation may turn to Enc and Adams (1992). They offer a well grounded discussion of the reasons why functional concepts do not supervene on the physical world and yet share with supervenient concepts the capability of defining regularities that physical concepts would overlook. Although their analysis was done independently of mine and came to my attention only after the present work was virtually complete, theirs complements mine by filling in some of the technical details of how a certain approach to behavioral concepts (which approach I believe we hold in common) can finesse a number of difficult philosophical issues relating to functional concepts, teleological explanations, and dispositions.

of responses to shape into an operant). There are few such principles that are valid across species, and in the case of human beings, many such regularities fail to generalize from individual to individual, or even from situation to situation involving the same individual.[8]

Critics of operant psychology are prone to dwell upon the fact that the latter type of operant regularity (a source principle) does not generalize well. As with evolutionary theory, however, the place where operant psychology arrives at broad generalizations is in the other type of regularity (consequence principles). This distinction may be used to explain why operant psychology is able to predict and control human behavior only in highly constrained situations. Prediction and control require not only that we have consequence principles telling us how various contingencies of reinforcement will affect operant behavior, but also that we have a rich supply of source principles telling us what physical circumstances will be reinforcing, what stimuli will be discriminative, and what operants will be available to reinforce. The more complex the organism and/or the environment, the less likely a given source principle is to be valid with respect to it. If on the other hand the organism is of limited cognitive complexity, there are only a few potential reinforcers in the environment, the conditions creating deprivation are predictable and subject to control, only a few prominent stimuli provide information about which manipulations of the environment will produce reinforcement, these stimuli are predictable and subject to control, and these manipulations are limited in number and within the behavioral repertoire of the subject, then we can predict and control the subject's behavior. For we can predict and control the sources of reinforcement, discriminative

---

[8]Killeen (1987) notes that Skinner defines each part of the standard three-term operant contingency functionally, but he defines the operations that establish these functional categories physicalistically--e.g., reducing a pigeon's access to grain causes access to grain to become reinforcing, delivery of grain to a pigeon immediately after a response having a certain consequence causes the class of responses with that consequence to become an operant, temporal contiguity between a stimulus and a contingency of reinforcement causes that stimulus to function discriminatively.

stimulation, and operant repertoire, and therefore, we can predict and control the input to the consequence principles that entail what the subject will do.

The Role of Source Principles. If we are going to analyze behavior experimentally, we must constrain the experimental situation in such a way that we reliably induce certain aspects of the environment/organism system to function in a certain way. For example, by maintaining a pigeon at 80% of its free feeding weight we guarantee that access to seeds will function as a reinforcer. By carefully shaping the pigeon's behavior, we guarantee that key pecks are operants. By simplifying the pigeon's environment so that it is free of distracting sights and sounds, we guarantee that illuminating the key with a red light will function as a discriminative stimulus. We can thereby insure that key pecks function as operants, that food delivery functions as a reinforcer, that red illumination on the key functions as a discriminative stimulus correlated with the fixed-interval schedule.

To do so we make use of certain facts, often quite specific to a certain species, about how to induce a certain functional relationship. Thus in the course of doing schedule research with pigeons, one does whatever is necessary to guarantee that the key peck will function as a discriminated operant throughout an experiment. This may seem easy to do, but as we learn more about pigeons we discover that (a) under some conditions a key peck may function as a respondent (e.g., when the key is repeatedly lit just before free delivery of food), (b) under other conditions it can become a superstitious response which is no longer under discriminative control (e.g., when concurrent schedules are not programmed with a changeover delay), (c) under other conditions it may function as an instinctive response (e.g., when a pigeon is trained to perform a complicated chain of responses which releases a pecking response), and (d) under still other conditions pecking of a somewhat different sort can function as an adjunctive response (e.g., when the primary reinforcer is food delivery on an intermittent schedule, and the pecking response is a form of schedule-induced behavior associated with consumption of water--cf., Falk, 1986).

To insure that behavior exhibits a certain set of functional relationships requires great skill. The range of experimental conditions under which we can confidently assert that a pigeon is engaging in purely operant behavior is limited. But operant behavior is real, and we know of no other way to study it rigorously. Since the inductive inferences from a behavioral experiment are projected to environment/organism systems that exhibit the same functional relationships, it is very important to be able to control the function of various aspects of behavior and the environment during an experiment. But this does not mean that the theory only applies to systems that are similarly constrained.

This is a subtle point, but an important one. The variables in our experiment are physical events, but they are not the physical events which induce certain aspects of the organism/environment system to function in a certain way. The latter events fall under such relatively uninteresting (i.e., highly restricted) regularities as: (1) Access to grain reinforces responding in pigeons that have not been fed, or (2) If the color of a key is correlated with a certain schedule of reinforcement, then that color can come to exert control over the pigeon's key pecks. Knowledge of such regularities is absolutely necessary for the success of the experimental program, and a detailed knowledge of them is one thing that distinguishes a competent animal psychologist from the rest of us. But such narrow regularities do not exemplify the theoretical principles of behavioral psychology. These theoretical principles attempt to state the ways in which certain physical parameters of functionally defined aspects of the organism/environment system affect one another. Such principles are limited in their ability to predict and control the behavior of complex organisms in complex environments. This limitation in the ability to predict and control does not, however, imply that the principles are not valid in complex environments.[9] At most, it simply implies that such

---

[9]The reasonable position on this issue is simply to say that it too early to tell (Staddon, 1983). The philosophical question at issue, however, is whether it is plausible for the operant research program to hold out the hope that its way of formulating behavioral principles might eventually lead to a valid, context-free account of human behavior. And in this respect, our analysis of operant psychology provides a strong case for answering yes, in the sense that there is reason to hope

principles are not useful for prediction and control of behavior in complex (i.e., natural) environments--something Skinner pointed out as early as 1938.

---

that the consequence principles of operant psychology will apply to human behavior in situations of arbitrary complexity.

CHAPTER SEVEN

A SOPHISTICATED REJOINDER BY PHILOSOPHERS

Many alleged solutions to philosophical problems have turned out to be little more than conceptual sleights of hand. Thus when a seemingly difficult problem becomes easy to solve if only we adjust our understanding of one or two key concepts, philosophers have learned to become suspicious. So instead of questioning whether behavior analysis can circumvent the types of counterexample that have bedeviled S-R psychology, philosophers are inclined to accept the line of argumentation developed in the previous three chapters, but to question whether behavior analysis (as thus represented) is truly behavioral (cf., Margolis, 1984, pp. 41-42; Rosenberg, 1988, pp. 57-65). They wonder, for example, how the meanings of behavioral concepts manage to coordinate so well with the meanings of mentalistic concepts, giving behavioral principles just the right spin to steer their way around the tangle of problems that folk psychology has thrown in their path. Are not the two sets of concepts logically connected somehow? But if they are logically connected, then behavioral categories are tainted with mentalism, and the principles making use of them are not significantly different from those of folk psychology. This, as noted earlier, is the form that sophisticated philosophical critiques of the Skinnerian program tend to take nowadays.

Up to this point, we (like Skinner himself) have focused upon distancing behavior analysis from S-R psychology. Perhaps the time has come to consider it done. The challenge now is to put some distance between behavior analysis and mentalism.

There are several ways to do this. One is to show that behavioral categories are not themselves a subspecies of mentalistic categories.

Assuming that a definitive feature of mentalistic categories is possession of propositional content, this could be accomplished by showing that behavioral categories do not have propositional content.[1] A second is to show not only that behavioral categories are not a subspecies of mental categories, but that they are not even logically connected with them. Although the current state of linguistics does not permit such assertions to be demonstrated decisively, it is sometimes possible to make a plausible case for them by showing that certain apparent logical connections break down under careful examination. And a third is to show that behavioral categories serve a different scientific purpose from that served by mentalistic ones.

Philosophers of a technical bent may consider this last item superfluous. If we know behavioral concepts are neither identical with, nor logically connected with, mentalistic concepts, this should suffice to quell the doubts about the behavioral credentials of behavior analysis. But even without such doubts, we would still wonder what all the fuss is about. What is so special about doing psychology behaviorally anyhow? Why this fastidious attempt to avoid mentalism? An answer to this question will not only bolster the defense of behavior analysis, but help us understand it. So let us begin at the end, and take the last question first.

I

We have already noted a kind of conceptual division of labor between behavioral and mentalistic concepts in our discussion of the example of the dog's tracking behavior at a fork in the road. We noted that the behavioral description tells us where the animal is making causal contact with its environment, and the mentalistic account explains how that contact comes about. The behavioral description, for example,

---

[1]Margolis (1984) notes that there are two quite different sorts of mental states--those such as beliefs and desires that have propositional content and those such as pleasures and pains that do not. It is only the first type of mental state that behavior analysis attempts to avoid reference to. This point receives further discussion in the following chapter.

identifies the absence of scent as the discriminative stimulus that controls the response of crossing to the other fork. This tells us something about the crossover response--it is not, for example, caused by the dog's hearing an animal running up the other fork, or by its seeing a path of matted grass leading from one fork to the other. But the behavioral description gives no account of how this discriminative stimulus leads to the response. On the ancient Stoic account, the dog knows that the animal it is chasing took one fork or the other, and it knows that the absence of scent indicates the animal did not take this fork, so it concludes that the animal must have taken the other fork. Thus, it performs a syllogistic inference. On an alternative account--viz., Thorndike's (and later, Skinner's) classical theory of instrumental (operant) conditioning--there is no underlying inference. The connection between discriminative stimulus and operant response is an unthinking habit reinforced by past successes. Obviously, the first account is in competition with the second. The behavioral description, however, is neutral between them. It simply tells us what stimulus caused the response of moving to the other fork.

What is the purpose of such a description? Behavior analysis aspires to arrive at a body of causal regularities it continually revises and refines. Behavior analysts hold the conviction (which is widely shared among natural scientists of an experimental, as opposed to theoretical, bent) that such a goal is easier to attain if it formulates causal relationships between properties that can be measured independently of one another. That way, if someone suggests that the amount of X is inversely proportional to the amount of Y, then we can arrange a procedure for measuring X and we can arrange another procedure for measuring Y, and we can check to see if the relationship is in fact what has been suggested. What we sometimes find is that for many quantities and under many circumstances, the suggestion is approximately true, but that for other quantities under other circumstances, it needs refinement. Given the independent measurability of X and Y, it should be possible to run some further experiments, and revise the suggestion to take these values and circumstances into account. The result is a more accurate formulation of the regularity, which formulation in turn becomes the target of further refinements, and so on.

This brings us to a distinctive purpose of behavioral descriptions. One important advantage of the categories of behavioral psychology over the categories of mentalistic psychology is that the quantitative variables that attach to behavioral entities are properties that can be measured independently of one another. The quantitative features of mentalistic entities, on the other hand, cannot be so measured.

The Conspiracy Theory. Let me emphasize right away that I am not claiming that discriminative stimulus, operant response, and reinforcing stimulus can themselves be defined independently of one another, nor that the presence of one can be confirmed independently of information about the others. In this respect, behavioral categories share a widely remarked feature with mentalistic categories. Rosenberg (1988) has described this interdependence of mentalistic categories as a "conspiracy" of desires, beliefs, and actions. Participants in a conspiracy do not act independently of one another, and true to the metaphor, one cannot establish the occurrence of a given desire, belief, or action independently of information (or assumptions) about the other two. A bodily movement establishes that the subject has performed a certain action only on the assumption that the subject holds certain beliefs and desires. Performance of a certain action establishes that the subject holds a certain belief or desire only upon the assumption that he holds certain other beliefs and desires. Thus, we never verify actions, beliefs, or desires independently of one another.

A similar conspiracy exists among operant-responses, discriminative-stimuli, and reinforcing-stimuli. We can infer that a certain bodily movement is an operant response only if we have evidence that its rate of occurrence is under the control of a reinforcing stimulus. We can infer that a stimulus is reinforcing only if we have evidence it controls the rate of responses upon which its occurrence is contingent. We can infer that a stimulus is discriminative only if we have evidence it controls the rate of an operant response by signaling its effectiveness at producing reinforcement. As Rosenberg (1988) notes, behavioral theory is in this respect so similar to folk psychology that it is difficult to see how it represents an improvement over it.

But the advantages of behavioral descriptions become apparent only when we turn to the quantitative aspects of behavior. Although both behavior analysis and folk psychology have the capacity to add quantitative measures to their basic categories, folk psychology's quantities inherit all the problems of the conspiracy of desires, beliefs, and actions, whereas behavioral psychology's quantities break free from the conspiracy of reinforcing stimuli, discriminative stimuli, and operant responses. This independence of behavioral quantities from one another is one important benefit of doing psychology the behavioral way.[2]

Quantifying Mentalistic Categories. Consider the quantification of belief, desire, and action. Intuitively, it makes sense to assign degrees of strength to these categories. Beliefs can be held so strongly that we speak of them as convictions. Others are mere reasoned judgments, and even less strongly held are speculations and hunches. Desires can burn, cool off, then become icy. We can be determined to perform an action, later become merely inclined towards it, then hold misgivings about it, and finally resolve never to perform it again. From such vague rankings of degree one can move to precise ordinal scales, and perhaps even to cardinal scales that assign real numbers to degree of belief, strength of desire, and inclination to act. A good deal of work has been done on this sort of thing, beginning with Bentham's utilitarian calculus, proceeding through the marginalist school of microeconomics, and extending to the current excitement about the Bayesian approach to rational choice. Much of this work addresses the problem of how we should act, and therefore is irrelevant to our discussion. But it is possible to interpret these theories as psychological models, and thus interpreted, they offer quantitative versions of folk psychology.

Like any theory, it has its strengths and weaknesses. I do not claim to know how these stack up, and I certainly am not in a position to assess its scientific potential. But I do see one problem with folk psychology that I do not see with behavior analysis. One cannot measure folk psychology's quantitative properties independently of one

---

[2]For a complementary account of the benefits of behavioral concepts see Enc & Adams (1992).

another. Consider, for example, how one might attempt to measure the difference between a weak belief and a strong one. Perhaps the most straightforward approach would be to infer the strength of a belief on the basis of the actions the subject performs. But a given act is evidence of the strength of a given belief only to the extent that we know (or can hold constant) the strength of the subject's desires and other relevant beliefs. Drinking the poison may be evidence of an increase in the strength of the belief that one's embezzlement is about to be discovered, but only if there is no reason to suspect a decrease in the desire to live following the death of a loved one, or an increase in the strength of the belief that one has a terminal disease, and so on.

The fact that dependent and independent variables cannot be quantified and measured independently of one another makes it more difficult to establish quantitative relationships experimentally, and then to refine our understanding of them. It once was assumed that scientific method itself required this kind of independence. Few philosophers nowadays, however, would claim that a science must feature independent measurability of cause and effect. If the history of science shows anything, it is that each discipline has its own epistemological peculiarities, and some well established sciences violate the principle of independent measurability. On the other hand, it is difficult to deny that independent measurability offers a possible epistemological advantage. The question is whether, in a particular science, such an advantage can be parlayed into sustained progress. And this question cannot be answered on philosophical grounds.

Quantifying Behavioral Categories. Consider the behavior of an animal on a type of schedule we have not yet discussed. On a so-called variable-ratio schedule, the delivery of the reinforcing stimulus is contingent upon the number of times the subject has performed the operant response, but this number varies from delivery to delivery. Over the course of ten or twelve deliveries, the number of responses required for reinforcement can be made to have a certain average, and this average can then be assigned to the schedule. A VR-5 schedule requires, on average, five responses for each delivery of reinforcement, a VR-10 requires ten, and so on. In general, VR schedules maintain high steady rates of responding.

How much difference, if any, does a change in ratio make in the rate of responding? This is a straightforward quantitative question with a straightforward quantitative answer. The number of operant responses required on average to earn a pellet can be varied through manipulation of the equipment. The cumulative recorder then gathers information about the effect (if any) of this change on the rate of responding. Basically, the marginal increase in rate of responding caused by a fixed increase in the VR average decreases as the ratio gets higher. A change from VR-5 to VR-10 causes a greater marginal increase in rate of responding than a change from VR-10 to VR-15, which in turn causes a greater marginal increase than a change from VR-15 to VR-20, and so on. At some point, x, an increase to VR-x causes no increase at all, and an increase beyond x causes responding to cease altogether. These findings can be summarized in the form of a function that maps the average value of a VR schedule onto the average rate of responding. The slope of this curve decreases as VR approaches x.

It has recently been discovered that this function is only approximately true--that it can be modified in a lawful way by other factors. Under the classical experimental protocol developed by Skinner, an animal is always kept at 80% of its free feeding weight. Thus, the number of pellets that can be earned during an experimental procedure is determined ahead of time, and the experiment ends when it has been earned. As ratios get higher, however, it is not always possible for the animal to earn its daily ration of food during the experimental session. When this happens, the animal gets a free snack after the experimental session of the pellets it did not earn. This arrangement is known as an open economy.

Its opposite is the closed economy. On this protocol, the animal gets no supplemental feeding. It eats only what it earns through its operant behavior. Under such conditions, the slope of the function relating VR averages to rate of responding changes significantly (Hursh, 1980, 1984). As ratios increase, the rate of responding increases also, and there is no point (short of physical exhaustion) at which the responding ceases altogether. This is a quite different relationship between contingencies of reinforcement and rates of responding than we get under the standard protocols of an open economy. What this discovery points out is the relevance of certain background conditions to the applicability of the earlier generalization about variable-ratio schedules. This is the sort of thing one expects with a progressive program of research that aims at formulating more and more accurate causal principles within its domain.

In brief, behavioral categories are designed to present their quantitative aspects in a manner that can be measured independently of the rest of the theory. This does not mean one does not need to make assumptions to measure (say) rate of responding or delay of reinforcement, but (except for those relating to the application of the behavioral categories themselves) these assumptions have nothing to do with psychology. They are assumptions, for example, about the operation of the equipment--that when the line on the cumulative record climbs higher this is because the rate of key pecking increased (and not because the armature went haywire), or that when the computer program was supposed to double the average length of delay to reinforcement, it actually did. Since these are not assumptions of the psychological theory being tested and revised, there is a sense in which these quantities are being measured directly--i.e., without the benefit of psychological inference.

Not everyone however is impressed. There is an oft repeated joke about a man who dropped his keys in his back yard one night, and went out front to look for them. When asked why, he replied that the light was better under the street lamp. This joke has been used more than once against behavioral psychology, which supposedly prefers to look for causal relations in the animal laboratory (where the light is good) rather than groping in the dark where the answers to the important questions are to be found. It is an amusing way to state an opinion, but considered as an argument, it begs the very question at issue. The premise of the joke is that the key is known to be in the back yard. We have no such information about the location of behavioral regularities. Behavior analysts have now been searching for such regularities for over half a century. We know there are regularities where it is searching. Furthermore, we know that behavior analysts have been able to refine and broaden their understanding of these regularities. No one knew ahead of time that fixed-interval schedules would produce scalloped cumulative records, or that the effect of increasing the ratio of a variable-ratio schedule would be greater under a closed economy than under an open economy, or that electrical shocks would be capable of maintaining behavior under fixed-interval schedules (as discussed below). These are genuine discoveries, and not trivial demonstrations of what we already knew or suspected. True, they are more like a collection of facts than a theory, but they are facts, and there is a certain coherence to the way in which they are accumulating. Furthermore, those who steep themselves in these facts begin to perceive relationships that most of the rest of us miss, and these relationships function in some ways like a theory. But finally, as we shall later see, a genuine theory of operant behavior has emerged, in a form Skinner neither anticipated nor approved of.

II

Despite (or perhaps because of) operant psychology's successes, the suspicion lingers that the behavioral concepts that appear in its principles are somehow equivalent to, or at least, logically linked to, mentalistic concepts. Let us turn to these allegations.

Are Behavioral Categories a Subspecies of Mentalistic Categories? Behavioral categories describe a system of functional relationships between the organism and the environment. An operant is not simply a response that the organism thinks will have a certain effect, it does have that effect. Thus, a key peck by definition compresses the surface of the key with a certain minimal force, a lever press by definition moves the lever through a certain arc. Similarly, a reinforcer is not simply a stimulus that the organism desires to occur. It is a stimulus that will alter the rate of behavior upon which its occurrence is contingent. And a discriminative stimulus is not simply a stimulus that has been correlated with a certain contingency in the organism's experience. It is one that successfully alters the organism's operant behavior with respect to that contingency.[3]

Beliefs and desires have propositional content. Fred believes that Neptune is the eighth planet from the sun, and desires that intelligent life be discovered on Mars. Fred's belief and desire are designated by the content of certain propositions introduced by the word *that*. Designations of discriminative stimuli and reinforcing stimuli, by contrast, do not accept *that*-clauses. Suppose, for example, that delivery of pellets functions as reinforcement for a rat's lever press. A mentalistic description might say, "The animal desires that a pellet should become available." A behavior analyst would not however describe this by saying, "The animal's lever presses are reinforced that a pellet become available." Instead, the proper description would be: "The animal's lever presses are reinforced by access to pellets." Instead of accepting a proposition as its object, the concept of reinforcement accepts an event or a state of affairs--such as access to pellets--as its object.

Consider the discriminative stimulus. Suppose that the sound of a buzzer functions as a discriminative stimulus marking the onset of a fixed-ratio 10 schedule. As soon as the buzzer sounds, a sequence of ten responses will result in the delivery of reinforcement. A mentalistic description might say, "The animal believes that the buzzer marks an opportunity to earn reinforcement by performing ten responses." A behavior analyst however would not attribute discriminative status to the buzzer by saying, "The sound of the buzzer is a discriminative stimulus that the fixed-ratio 10 schedule has begun." Instead of attributing a content to the stimulus, the behavior analyst will attribute a causal function to it, as in: "The sound of the buzzer signals the onset of the fixed-ratio 10 schedule." This tells us that the buzzer functions for the animal as a means of contact with the schedule. It attributes an effect to the stimulus, but not a content.

Analytic philosophers have noted that mentalistic statements create opaque contexts within which substitutability of identicals fails. Neptune, for example, is identical with the planet whose orbit was

---

[3]The account of behavior afforded by causal principles making use of such concepts relates an animal's behavior to its environment. Accounts making use of mentalistic categories may be intrinsically incapable of doing so (cf. Fodor, 1980).

predicted by Leverrier and Adams. The principle of substitutability of identicals tells us that we should be able to substitute "the planet whose orbit was predicted by Leverrier and Adams" for "Neptune" in a sentence without affecting its truth or falsity. Thus, "Neptune is the eighth planet" (by the principle of substitutability of identicals) logically entails "The planet whose orbit was predicted by Leverrier and Adams is the eighth planet." This principle does not, however, apply within the clauses that specify the propositional content of beliefs and desires. For example, if Fred believes that Neptune is the eighth planet, this does not imply Fred believes that the planet whose orbit was predicted by Leverrier and Adams is the eighth planet. And although Mars is identical with the fourth planet, Fred's desire that intelligent life be discovered on Mars does not imply (by substitutability of identicals) that Fred desires that intelligent life be discovered on the fourth planet.

Behavioral categories, on the other hand, create logically transparent contexts. If the sound of the buzzer is functioning as a discriminative stimulus marking the onset of a fixed-ratio 10 schedule, and a fixed-ratio 10 schedule is identical with the schedule most widely used in operant laboratories, then (by substitutivity of identicals) the sound of the buzzer is functioning as a discriminative stimulus marking the onset of the schedule most widely used in operant laboratories. And if the animal's lever presses are reinforced by access to pellets, and access to pellets is the customary reinforcer used with rats, then (by substitutivity of identicals) the animal's lever presses are reinforced by the customary reinforcer used with rats. These examples suggest that behavioral categories are not a subspecies of mentalistic categories.

Are Behavioral Categories Logically Connected to Mentalistic Categories? Behavioral descriptions may not be a subspecies of mentalistic concepts, but one may still wonder if the two are not somehow logically connected with one another. Otherwise, how can behavioral concepts share so many logical implications with mentalistic ones--especially the implications that permit them to sidestep the counter-examples of folk psychology?

The brief answer to this question is that behavioral concepts are functional. Their meaning is tied to the causal role of parts of the organism/environment system. When a shift in belief or desire disrupts that system to the point that some part of it changes its causal role, then the shift in mental state is accompanied by a shift in behavioral state. This is not, however, because there is a logical connection between mentalistic and behavioral descriptions. The connection is a nomic one. According to folk psychology, beliefs and desires are capable of altering the relationship of the organism to the environment. When this altered relationship changes the causal role of some aspect of the environment/organism system, then the behavioral description of the organism changes as well. This is not, however, because there is a logical implication that if a certain mental state changes, then so does the function of certain aspects of the behavior/environment system. Changes in mental state may or may not be accompanied by changes in behavioral state, depending upon the way the changed state interacts with the other mental states of the organism.

There is however a certain plausibility to the claim there is a logical connection between the two sets of concepts. It appears on first glance, for example, that if a stimulus is desired then it is a (positive) reinforcer, and if it is a (positive) reinforcer then it is desired. Neither half of this relationship, however, withstands careful scrutiny. One of the differences between a desire and a reinforcer is that a subject can desire something that does not exist. Ponce de Leon had a desire to find the fountain of youth, but there is none. Countless inventors had a desire to find a source of perpetual motion, but it cannot be found. These non-existent objects of desire cannot have been reinforcers. A reinforcer is part of an organism's environment, something it has causal contact with.

Let us look at the opposite implication. If a stimulus is reinforcing, is it not desired? This may seem incontestable. Higher cognitive capacities may make it possible to desire things that are not reinforcers, but surely anything that is a reinforcer is desired. If access to food and water is reinforcing, then surely it must be desired. What better evidence of its being desired could there be than its power to reinforce? Even this inference, however, is questionable. One of the most puzzling, yet well established, results of operant psychology is the ability of electric shocks to maintain responding in the context of a fixed-interval schedule (Morse & Kelleher, 1977). This phenomenon has been confirmed for a variety of organisms (including monkeys, pigeons, and rats) in a variety of laboratories. On other schedules, electrical shocks function as expected--i.e., they decrease the rate of behavior upon which they are contingent. But on a fixed-interval schedule that gives an animal an opportunity to deliver a shock to itself by performing a response (pressing a lever, pecking a key, etc.) after the elapse of a certain fixed-interval of time, the animal's rate of lever pressing generates the usual fixed-interval scallop on a cumulative record. The animal could easily avoid shocks. All it has to do is fail to press the lever. It clearly is capable of doing so, because on other schedules it does. But on a fixed-interval schedule, electric shocks maintain behavior.

It is not clear why this happens, but the most straightforward interpretation is that electric shocks are functioning as reinforcers (although see Pitts & Malagodi, 1991, for evidence that something else is going on to maintain responding). It is difficult however to believe that the animal wants to receive a shock. Otherwise, why would it avoid shocks under all other circumstances? On the other hand, one could ask, Why would the animal continue to press the lever, thereby delivering itself shocks, unless it wanted to receive shocks? Yet why would it want to receive a shock? There is a problem for the mentalistic description no matter what desires we attribute to the animal. And it is this problem itself that indicates the absence of a logical implication. If being a reinforcer logically implied being an object of desire, then the assumption that shocks function as reinforcers would imply they are desired. But we hesitate to make this inference. Why? Because the ability of electrical shocks to maintain the rate of responding supports the contention that they are reinforcers in a way that it does not support the contention they are desired. This indicates an absence of logical implication between the concept of being a reinforcer and the concept of being desired.

Similar considerations suggest a lack of logical connection between the concept of being a discriminative stimulus and the concept of being believed. First, consider the question of whether certain types of belief imply a discriminative function for a stimulus. Suppose Fred and Tom play racquetball often, and Fred has formed the opinion that when Tom drops the ball far from his body on his serve, he will attempt a passing serve down the right-hand wall. Fred believes that on this type of serve, an immediate dash to the back corner positions him well for a winning return, which is the desired outcome. Does this imply that Tom's dropping the ball far from his body on his serve functions for Fred as a discriminative stimulus that the response of dashing to the back corner will be effective in bringing about reinforcement? Not necessarily. Suppose, for example, that Fred formed this opinion by reading a book about racquetball tactics. Beliefs issuing from such sources are notoriously ineffective at modifying behavior. There are a whole series of reasons why. To begin with, Fred may not be able to distinguish between balls dropped far enough away to set up a passing serve and those not. Furthermore, Fred may not be able consistently to hit winning shots from the back corner. And Tom may be able to hit a variety of serves off a ball dropped far from his body. Dashing to the back right corner may not be an effective strategy for some of these. Finally, under the circumstances of a real game, Fred may simply be incapable of connecting the right response to the stimulus. Things happen so fast and there are so many things to think about, he may not be capable of doing what he knows to be the right thing. So there are at least four different reasons why the relevant stimulus may not function discriminatively in the appropriate way, even though Fred may hold the belief that one might expect to underlie such a function. This supports the contention that there is no logical implication from our beliefs to the ability of certain stimuli to function discriminatively.

Now consider the possibility of an implication in the opposite direction. Suppose the sound of a buzzer functions for an animal as a discriminative stimulus marking the onset of a fixed-ratio 10 schedule. Does this imply that the animal believes that the sound of the buzzer signals that a sequence of ten operant responses will be reinforced? Well, not necessarily. Some of the animals that can acquire a discriminated operant probably do not have the concept of the number ten, and may not have very many other concepts either. One can even question whether rats and pigeons (both of which readily acquire discriminated operants) have any beliefs at all. And here again, nothing in my argument depends upon whether they do or do not. The mere fact that there is a genuine question about whether they have beliefs is itself a point in my favor. If we know that an animal has a discriminated operant but do not know if it has any beliefs, this implies that the concept of having a discriminated operant and the concept of a having a belief are non-equivalent. Even among human beings (the animal that has beliefs if any does), the case for non-equivalence is close to conclusive.

The Costs of Conceptual Independence. Critics have often claimed that behavioral theories are either obviously false or else non-behavioral. The continued (albeit modest) expansion and success of the behavioral program shows they are not obviously false, and our preceding argument shows they are not non-behavioral. So we have passed between the horns of the dilemma. I would count this a success. It comes, however, with a cost. This cost is one I should think behavior analysts should be glad to pay, but it is one that should be explicitly stated.

To explain this cost, we need to make a few preliminary points. First, let us distinguish between a causal description, and a causal explanation that is supported by a lawlike regularity. We sometimes are in a position to say what the cause of some event is, but at the same time would be at a loss to state a valid causal regularity that connects cause to effect. When we describe a stimulus as a reinforcer, we are attributing a causal power to it. Although we are sometimes wrong about such attributions, and though we probably fail to notice important reinforcers more often than we think, just knowing that a certain response is being maintained by a certain reinforcer is not in itself a distinctively scientific explanation of the behavior. It does not subsume the response under a causal regularity. It simply identifies the part of the environment that causes (or is part of the cause of) the behavior. This does provide an explanation of sorts. It identifies a cause, and the identification of a cause does explain. But this is not the distinctive mode of causal explanation offered by science, where cause is connected to effect by means of subsumption under a valid causal regularity (typically quantitative).

As post-positivistic philosophers of science are well aware, however, subsumption of a cause under a causal regularity is not the only (or even the most important) mode of scientific explanation (Cummins, 1983a, 1983b). There is at least one other type of scientific explanation, in which we explain how one thing is able to cause another by analyzing the process that underlies the causal relation. Radiation causes cancer by damaging strands of DNA. Standing water causes an increase in malaria by abetting the reproduction of mosquitoes that spread the disease. Subsumptive explanation sharpens our knowledge of causal relations by giving us a sense of proportion about the quantitative aspects of cause and effect, the second deepens our knowledge by giving us a sense of what mediates the causal relation.

Behavioral descriptions are supposed to be a prelude to the formulation of the causal regularities that make subsumptive explanation possible. Mentalistic descriptions are preludes to the second type of explanation. Whether either type of description succeeds in attaining its explanatory purpose or not is beside the point, which is this: the logical gap between mental and behavioral concepts has, as a corollary, the possibility that mental concepts can offer a type of explanation that behavioral concepts cannot offer. This is a possibility I think behavior analysts should accept with equanimity, but it is one Skinner fought tooth and nail. Such is the challenge taken up by radical behaviorism, to which topic we now turn.

PART THREE

WEIGHING THE STRENGTHS AND WEAKNESSES OF RADICAL BEHAVIORISM

> I should not want to try to *prove* that there are no innate rules of grammar or internal problem-solving strategies or inner record-keeping processes. I am simply saying that an account of the facts does not require entities of that sort, that we do not directly observe them introspectively, and that an alternative analysis is more likely to be successful in the long run. (Skinner, 1984b, p. 663)

Radical behaviorism is Skinner's ontological position with respect to psychology. Radical behaviorism is distinct from the science of behavior, and is distinct from the methods and research strategy associated with this science (although behavioral psychologists sometimes mistakenly imply otherwise, as when Davey, 1981, writes as if radical behaviorism is identical with Skinner's non-reductive approach to psychology). It is easy to confuse the three, however, because the philosophy supports and defends certain attitudes associated with the science of behavior and its methods. Not that every behavior analyst is a radical behaviorist, but radical behaviorism is the most widely acknowledged attempt to give certain widely held attitudes a philosophical justification.

Perhaps the most important of these attitudes is the opinion that behavior analysis can give a complete account of behavior. Exactly what type of completeness is thereby implied is itself a difficult question. But the attitude itself is real enough and it affects scientific behavior, discouraging curiosity about other forms of research and contributing to the isolation of the program. Correlative with this attitude is a distrust of abstract theories, especially those about underlying processes. Some behavior analysts feel that mentalistic or cognitive theories are "dangerous" because they "seduce" behavior analysts, who sometimes "fall off the deep end" when looking into them.[1]

Behavior analysis may offer fewer opportunities for the exercise of imagination than do depth psychology or cognitive theory, with their repressed memories and defense mechanisms, or their information processing structures and long-term memory stores. But the science of behavior can hold out the hope of discoveries that could save modern civilization from the threats of pollution, overpopulation, and nuclear war. This is not to say that such discoveries are inherently beyond the grasp of alternative methods, but behavior analysis takes a more direct route to knowledge of the causes of human behavior (Skinner, 1967), and under the currently desperate state of things, "we [cannot] spare the time to worry about internal states as models" (Skinner, 1984b, p. 664).

What we think of as the *philosophy* of radical behaviorism is the set of doctrines by which Skinner has defended these attitudes. Slowly, these attitudes are changing. Some behavior analysts now acknowledge that cognitive theory is useful in accounting for certain behavioral phenomena, and they question whether behavior analysis can provide a complete account of behavior. Their theories are increasingly abstract and mathematical; and some view Skinner's conviction that they are about to discover solutions to difficult social problems as more a source of embarrassment than an inspiration. But I am getting ahead of myself. First let us examine radical behaviorism itself, and save for later a discussion of what is taking its place and what this implies about changes in the prevailing climate of opinion and attitude.

---

[1]These are all words or phrases that I have heard behavior analysts use in describing mentalistic psychology.

CHAPTER EIGHT

WHAT IS RADICAL BEHAVIORISM?

Willard Day (1987) writes that the earliest published use of the term *radical behaviorism* by Skinner appeared in the "Rejoinders and Second Thoughts" section of his 1945 paper, "The Operational Analysis of Psychological Terms."

> The distinction between public and private is by no means the same as that between physical and mental. That is why methodological behaviorism (which adopts the first) is very different from radical behaviorism (which lops off the latter term in the second). The result is that while the radical behaviorist may in some cases consider private events (inferentially, perhaps, but none the less meaningfully), the methodological operationist has maneuvered himself into a position where he cannot. (Skinner, 1945, p. 285)

This passage distinguishes radical behaviorism from the view that science can study only public (intersubjective) events. Instead of contrasting public with private, and then limiting science to the study of the public, Skinner contrasts physical with mental, and then eliminates the mental. In other words, "radical behaviorism argues that there are no such things as *mental* events" (Day, 1987, p. 19). Skinner usually contrasts this with methodological behaviorism, which argues that mental events exist, but cannot be studied scientifically.

The Abbreviated Defense. Skinner's typical exposition of radical behaviorism spins outward so quickly that we easily lose track of its basic claim. He no more than states that mental entities do not exist than he rebuts Everyman's first objection to this claim, which is: I know the mind is real because I observe it introspectively. Skinner's reply is that no, you do not observe the mind introspectively, you just observe certain inner events. Your feelings undeniably are real, but the desires you would infer from them are not. Your inner speech likewise is real, but the beliefs you would infer from them are not. Your subjective images obviously are real, but the intentions, plans, and expectations you would infer from them are not.

Skinner does not have too much more to say on this point, but his position is clear enough. The inner world we subjectively observe can, for purposes of science, be interpreted as a world of events. These events presumably enter into causal relations with other events, but there is no necessity that we interpret these events as having propositional content. Beliefs, desires, intentions, etc., have propositional content but are not objects of subjective awareness; whereas inner speech, feelings, images, etc., are objects of subjective awareness but do not have propositional content. We are directly aware of the subjective realm but we infer the mental realm. Therefore, Everyman's proof of the reality of the mental realm is ineffective. So much for naive mental realism.

Quickly, Skinner moves to the next pressing issue. Granted we are not directly aware of the mental realm, but is it not reasonable to infer its existence to make sense of our subjective experiences? Skinner's answer is no, we can interpret our subjective experiences behaviorally. Some subjective phenomena are responses, and can be analyzed as operants or as respondents. Others function as reinforcers, discriminative stimuli or conditioned stimuli.

One might reply that such a behavioral interpretation is inadequate compared to the mentalistic interpretation. Skinner's rejoinder to this is that, as a matter of fact, private events are so difficult to describe accurately that very little meaningful progress can be made on the question of how to interpret them. His argument for this position was first presented in the 1945 paper in which he introduced the term radical behaviorism. In it Skinner asks how he might establish a verbal practice denoting some subjective event. Since the event is not publicly observable, he will not be able to know whether another person is having an inner experience of the same kind unless there are some public accompaniments of the event that can be used to corroborate its presence. In this way, he can help another person learn to form a discrimination of that type of event. Unfortunately, he is not in a position to know whether the public events correlated with his subjective experience is also correlated with the other's. This is not a matter of principle, for Skinner, but one of degree. The circumstances under which we learn the vocabulary for describing private events cannot be relied upon to produce a uniform usage. He is not saying we cannot communicate about the subjective realm, but that we cannot do nearly as good a job of it as we can with the public realm. As a result, the objects of subjective awareness do not constitute a promising domain for basic research.

Skinner's argument has two targets. The first is the introspectionist psychology that dominated Harvard at the time Skinner entered graduate school. The introspectionists never arrived at agreement on the basic facts of consciousness, and Skinner has an explanation why: the way we learn the vocabulary of introspective terms renders the reference of that vocabulary less determinate than the vocabulary of public events. The second target is the opinion that only events observed by two or more people can be legitimately discussed by science (Skinner's way of putting the characteristic thesis of methodological behaviorism).

Skinner replaces these qualitative epistemological distinctions with distinctions of degree. Instead of saying that psychology must in principle start with our knowledge of the inner world, or that psychology cannot in principle ever talk about the inner world, Skinner says science should not give the events in the inner world much evidential weight. He leaves open the possibility that the inner world can at least be interpreted scientifically, but questions whether there can be a useful experimental program based on it. Thus, despite the fact that the inner world is real, it cannot provide reliable evidence of the existence of mind, even inferentially. Any such evidence will have to come by way of public events. But this is terrain upon which Skinner expects his non-mentalistic approach to hold its own. The point of his discussion of the vocabulary describing inner events is to neutralize the argument that we need a realm of entities having propositional content if we are to make sense of our inner experience. Our inner experience is not determinate enough to bear that much weight.

This is the abbreviated exposition of Skinner's radical behaviorism.

It consists of a bald assertion that mental states do not exist, together with a brief defense of that assertion against two obvious objections. His discussion does not resolve the issue, but that is not the point, which is to move the debate onto Skinner's home turf. The issue between behaviorists and mentalists is "primarily empirical rather than logical" (Skinner, 1972, p. 555) and must be resolved by the experimental analysis of public behavior.

The Tendency to Shift the Meaning of Radical Behaviorism. Day (1987) writes that "in speaking of his own professional views and interests," Skinner was more likely to speak of "'the analysis of behavior', or 'operant analysis', or 'a science of behavior'" than of radical behaviorism (p. 20). Presumably, this is because Skinner meant something rather specific by radical behaviorism--something not communicated by these other terms. But if Skinner used the term narrowly and infrequently, his defenders and critics alike have tended to use the term broadly and often. The most common expansion employs radical behaviorism as a general term to characterize Skinner's entire program of research (e.g., Davey, 1981; Leahey, 1987). Some would go further, however, and use it to refer to Skinner's general point of view--not simply on psychology but even on social reform, religion, Western civilization, etc. And at least one scholar has defined radical behaviorism as "the effect that Skinner's thought happens to have on the behavior of people" (Day, 1983, p. 101). Unfortunately, such expansions would surrender a distinctive technical term that Skinner found to be valuable in a limited context, and (re)define it in a way that either makes it synonymous with terms that are already available, or else renders it so vague as to be useless. This is not the way to analyze someone else's thought, so we shall interpret *radical behaviorism* narrowly, as Skinner intended.

Relation to Mind-Body Identity Theory. There are several philosophical misunderstandings about radical behaviorism that need to be removed before we examine Skinner's extended defense of it. The first has to do with the traditional philosophical position known as the Mind-Body Identity Theory. According to this theory, subjective states such as pleasure or pain are interpreted as identical with physiological states--usually brain states. This position is vulnerable to philosophical attack (for reasons similar to, if not identical with, those that render the physicalism of S-R psychology vulnerable), so it is important to know whether radical behaviorism is committed to it. If it is, then we could quickly dispense with it, and save ourselves some trouble.

Philosophers sometimes think Skinner is committed to this theory because of passages such as the following:

> Mentalism kept attention away from the external antecedent events which might have explained behavior, by seeming to supply an alternative explanation. Methodological behaviorism did just the reverse: by dealing exclusively with external antecedent events it turned attention away from self-observation and self-knowledge. Radical behaviorism restores some kind of balance. It does not insist upon truth by agreement and can therefore consider events taking place within the skin. It does not call these events unobservable, and it does not dismiss them as subjective. It simply questions the nature of the object observed and the reliability of the observations.
>
> The position can be stated as follows: What is felt or introspectively observed is not some nonphysical world of consciousness, mind, or mental life but the observer's own body. (Skinner, 1974, p. 17)

This may seem a straightforward statement of the Mind-Body Identity Thesis, but things are more complex than they seem.

Skinner's autobiographical reflections on the birth of radical behaviorism include the following passage:

> I was not concerned with the nature of the stuff of which the mind was composed: "The behavioristic argument is not that of the naive materialist who asserts that 'thought is a property of matter in motion,' nor is it the assertion of the identity of thought or conscious states with material [brain] states." (Skinner, 1979, p. 117)

The sections enclosed in quotation marks are from *A Sketch for an Epistemology*, an unpublished rehearsal of radical behaviorism written in 1932. In them, Skinner asserts that nothing in his position implies or presupposes that subjective states are identical with brain states. It may seem that he can be pushed to defend this position, however, for he says that subjective states are states of our bodies. How are we to resolve this apparent contradiction?

The key lies in Skinner's analysis of the fundamental terms of behavior analysis. When he says that subjective states are states of our bodies, he means only that these are states of our bodies in the same sense that stimuli and responses are states of our bodies--i.e., states that are functionally defined. This means they will not be individuated as brain states--on the basis of their physiological properties--any more than a generic stimulus or response would be. And because the aspect of the Mind-Body Identity Theory that makes it so vulnerable to counter-example is precisely the identification of subjective states with physiologically individuated states, there is reason to believe Skinner can avoid the philosophical problems of materialism. In any event, radical behaviorism is not the same as, nor does it imply, Mind-Body Identity. Therefore, no quick and easy refutation of radical behaviorism is forthcoming on this front.

The Relation to Logical Behaviorism. Another position that commentators often attribute to Skinner is the doctrine of logical (or philosophical or analytical) behaviorism, according to which all mentalistic terms can be defined behaviorally. Even Skinner's close friend, philosopher W. V. O. Quine, says (somewhere) that he could never convince Skinner to give up this doctrine. Actually, the truth would seem to be that Skinner never held the doctrine--not at least in the sense that philosophers define it.

The basic problem of interpretation is due to the failure to distinguish between subjective terms and mentalistic terms. Philosophers sometimes blur this distinction, but Skinner's first line of defense of radical behaviorism depends upon drawing this distinction sharply. The aspect of logical behaviorism that is vulnerable to philosophical criticism is the thesis that mentalistic (not subjective) terms are subject to behavioral definition. Skinner does not wish to address this issue, because he thinks mental terms have no reference anyway. All he wants to argue is (a) that subjective events can be interpreted behaviorally, in the sense that they can be analyzed as referring to inner stimuli and responses, and (b) that mentalistic interpretations of such events are thereby rendered superfluous. He is not trying to *define* mentalistic terms, he is trying to give an alternative behavioral interpretation of the subjective events that Everyman thinks offer the best evidence of the mental realm. Here again, the only point of Skinner's philosophical maneuver is to shift the debate about the existence of the mental realm to the behavioral laboratory. For this purpose, he does not need to endorse logical behaviorism.

The Troublesome Relation to Methodological Behaviorism. There is one final confusion about radical behaviorism that we need to address before turning to Skinner's experimental defense of it, and this is the question of whether radical behaviorism turns out (when you use philosophical analysis to peel away the superficial layers that indicate otherwise) to be identical with methodological behaviorism. Given the frequency with which Skinner contrasts radical behaviorism with methodological behaviorism, one might think that methodological behaviorism would be the last thing one could confuse radical behaviorism with. But philosophers love to show that something is really its opposite, and so some have argued that when you push radical behaviorism hard enough, it becomes indistinguishable from methodological behaviorism. It takes a bit of work to motivate this view, so let us back up a step.

On our interpretation of radical behaviorism, it stands in exactly the same relation to the philosophy of mind as atheism stands to theology. Now atheism is not such an easy position to defend. There is an inherent difficulty with any attempt to deny the existence of something which itself is difficult to characterize in a positive way. An atheist almost by necessity must operate within a cultural context that provides a conception of the God whose existence is to be denied. In our culture this is the God of Judaism, Christianity, and Islam--a God who performs miracles, reveals moral truths, makes agreements, and responds to requests for help. This God spoke to Moses at Mount Sinai and made a covenant with the ancient Hebrews, then He may or may not have later sent a Messiah to institute a new covenant and a new moral order, and still later He may or may not have spoken through the prophet Mohammed to reveal a plan for an even more complete moral order.

Arguments against the existence of such a God are not likely to join issue with every affirmation of the existence of God--not even within our own culture. The Deist, for example, posits the existence of God. But this God performs no miracles, answers no prayers, delivers no revelations, makes no agreements. This God creates the universe, imposes order upon it, then allows it to proceed according to natural law. Although Deists affirm their view of the universe by asserting the existence of this God, atheists may find little to dispute in this affirmation except the use of a theistic vocabulary to express it. Western atheists have a genuine disagreement with Judaism, Christianity, and Islam; but with Deism they have little more than a quibble.[1]

At the Rice Symposium, Skinner got into a quibble with Michael Scriven, and this quibble is the basis (and so far as I can tell, the only basis) for the extraordinary view that radical behaviorism is not significantly different from methodological behaviorism. The interchange begins with a question from Scriven.

> Could not many of Professor Skinner's remarks be construed "as a procedure for giving a sympathetic analysis of mentalistic and psychic explanations?" Did he not "show that these [explanations]

---

[1]The theological analogy can be carried further. We have seen that Skinner's use of traditional behavioristic terms such as *reflex*, *stimulus*, and *response*, marks a significant departure from the usage of S-R psychologists. On the one hand, Skinner is denying the existence of mind, but on the other he is expanding the meaning of S-R vocabulary to include some of the phenomena that had previously been described mentalistically. This is similar to the theological move of the atheist who denies the existence of God, while simultaneously expanding the scope of non-religious concepts to include phenomena that previously had been described only in religious terms.

This leads to a debate between religious liberals on the one hand--for example, the Unitarians (within Christianity) and the Reconstructionists (within Judaism)--and atheists (e.g., secular humanists such as Paul Kurtz) on the other. The religious liberals want to continue to use words such as *God*, *grace*, *faith*, etc., but to alter their meanings significantly from the orthodox ones. The secular humanists, on the other hand, would describe the same phenomena (to the extent there is agreement on the facts between the two groups--and to a surprising degree, there is) with non-religious concepts such as *ultimate principles*, *sense of well being*, *basic trust*, etc.

This is not to imply that there are no substantive differences between the two. There are. It is only to say that unless one understands the way certain conceptual moves are used to gain leverage in such disputes, one easily misses the point.

> can be legitimately construed as references to a state of the organism in which all of the dispositions do produce behavior of certain kinds?" Does not the analysis argue that these states ought, properly, be referred to in behavioral terms and--"a crucial point"--should a state induced by an earlier exposure persist in an organism and modify later behavior, it must be referred to in these terms?

Skinner's response is:

> "I do feel, in a way, that I am offering a reinterpretation of a mentalistic analysis--that you can redefine if you like--but that [redefinition] is always a dangerous kind of thing...." It is preferable "to use terms which come out of an analysis" rather than apply terms from some other source.
>
> There exists, at present, a gap between terminal events in our behavioral analysis. "Mentalistic explanations, physiological explanations, and conceptual inner events as explanations" are all "on a par" in their attempts to fill this gap. Some day the gap will be filled. The conceptual formulation is not helpful and the "mental properties added to the conceptual are a distraction." It is most likely that the physiological explanation will "win out." Whatever it will be, it seems "reasonable" to carry out the "original decision" to "get on with the functional analysis of terminal events," rather than wait for the gap to be filled. (Wann, 1964, pp. 103-104.)

Other questions and answers follow, including one by Skinner which Sigmund Koch characterizes in the discussion as "an intolerant answer" (p. 105).

It is in this context, in which Skinner would be inclined to make efforts to demonstrate his tolerance, that Scriven presses his point home.

> To follow up on the answer given to Scriven's earlier question, it appears to be the case that (1) mentalistic concepts can legitimately be interpreted as referring to a state of the organism which can alter future dispositions to behave and which can be explained in terms of earlier reinforcement schedules and (2) introspection is allowed in the sense that there are some parts of the universe to which an individual has direct access but to which no other individual has direct access.
>
> "Can't we put these two together and say that organisms are sometimes able to detect their own states in a way which others are

> not able to detect and that, moreover, these states which they detect can also be regarded as giving a guide to future behavior, though not a perfect one because of . . . the language community's inadequacies in establishing constancy of labeling here? . . . Where have we got with a behaviorism that allows us on the one hand a legitimate interpretation of mentalistic language and [on the other] a direct access to 'mental' states, with 'mental' here meaning . . . a state of the organism which can be directly perceived and which . . . is an indicator of subsequent changes in behavior? . . . Why do you feel that you are, in fact, still in some sense a radical behaviorist rather than someone who is making an extremely useful recommendation about the way in which we should prune the surplus out of mentalistic language?" (Wann, 1964, pp. 105-106.)

Skinner responds with the following passage:

> "I am a radical behaviorist simply in the sense that I find no place in the formulation for anything which is mental." This is a minor issue, the major reason for his position being his certainty that the reports about the "internal states" are not adequate. "They [the internal states] exist--we can create a vocabulary for talking about them and part of human progress has been the improvement of our description of these things." But this does not increase our "introspective clarity," rather, it helps us "understand the relevance of forces in our lives and in our history and in the current environment." If "behaviorism" means "simply the issue of the stuff of which the mental event is composed" then he is a radical behaviorist. Otherwise, he is a "methodological one, arguing [that] there are better ways of formulating relations than by setting up so-called intervening variables." (Skinner's comments as paraphrased by T. W. Wann in Wann, 1964, p. 106)

This passage poses a challenge for our interpretation of radical behaviorism. How can it be a minor issue whether there is a place in the formulation for anything mental? And how can Skinner's behaviorism be identical with methodological behaviorism, instead of contrasting with it?

What are we to make of Skinner's response? Day (1987) speculates that "Skinner's use of the expression 'radical behaviorist' in this circumstance might be taken to show merely intra-verbal control from Scriven's question" (p. 20), which is Day's quaint way of saying that

Skinner's reply is not necessarily representative of his reflected position. I would agree. I think Day has also located the source of the problem--namely, Scriven's question. No orthodox mentalist would be willing to use the term *mental state* to refer to the behavioral dispositions that Skinner suggests be used in place of ordinary mental concepts. For Skinner is speaking only of dispositions associated with subjective states--our feelings, inner speech, images, etc. These are not identical with the desires, beliefs, and intentions to which an orthodox mentalist wants to refer. But Scriven's question begins by proposing that we take mentalism to be identical with the position that the mind consists of these subjective events, plus the dispositions that accompany them. Given this remarkable proposal, arguing with Scriven about radical behaviorism is as productive as arguing with a Deist about atheism. One has lost contact with the issues that were worth arguing about, and begins to talk about how best to talk, which is exactly what Skinner does.

As one might expect, given the context, he does not do a particularly good job of it. First, Skinner says his inability to find a place for anything mental "is a minor issue." Well now, it is a minor issue only for those willing to use the term *mental* as Scriven would reconstruct it, which certainly would not include any orthodox mentalist. But Scriven's maneuver has thrown Skinner off balance. Why? Because the major difference between radical behaviorism and methodological behaviorism has been erased, so Skinner now uses the term radical behaviorism to refer that part of his philosophy which is left over when mental entities cease to be interpreted as inner states having propositional content--which is to say, not much. If everyone had been willing to limit mentalism in the way Scriven is, there would have been no need for radical behaviorism. But of course Scriven is unrepresentative of mentalism's core tradition, which posits not only subjective events, but also beliefs, desires, intentions, and a wide array of mechanisms for storing, altering and manipulating these entities.

Radical behaviorism is the thesis that none of these inferred states and mechanisms exists. Notice that we have yet to say the first word about Skinner's empirical defense of this position. What we have done is simply to preclude any misunderstandings about the position he intends to defend, and to summarize the arguments by which he deftly pushes the debate out of the philosopher's study and into the psychologist's laboratory. For despite the fact that Skinner rarely draws the connection explicitly, the principal support for radical behaviorism comes from his own program of research.

CHAPTER NINE

THE SCIENTIFIC CASE FOR RADICAL BEHAVIORISM

Philosophers typically regard behaviorism as either a logical (analytical) position or a methodological position. The former holds that mental events *can* be analyzed as behavioral dispositions, the latter that they *should* be (and that those which cannot so be analyzed should not receive scientific treatment). An element common to both positions is a refusal to countenance inferred entities. Radical behaviorism, with its sweeping denial of the mental realm, might be expected to do the same, but it does not. Skinner is a philosophical realist (Meehl, 1986); he has no objection to inferred entities, so long as the evidence warrants. His objection to mentalism is not that it posits entities we cannot directly confirm (as we shall see, he posits some inferred entities himself), but that the evidence does not warrant mentalism's type of inferred entity. There is no brief way to explain Skinner's reasons for saying so. His argument gathers force with the development of his experimental program.

I

Skinner's first opponent is common sense. The ordinary person interprets intentional behavior as due to an inner agent. Increased experimental control over behavior reduces the need to posit an inner agent. Projecting the trend forward, Skinner suggests the need will disappear altogether. Hence, the progress of an experimental program that discovers environment-to-behavior regularities obviates the need to posit an inner agent.

One sees this argument taking shape in Skinner's earliest research.

In the spring of 1930 Skinner studied what he called the eating reflex of the white rat. To do this he deprived the rat of food for twenty-four hours, then put it in a box containing a food bin that dispensed a pellet when the rat opened the door to it. He found that "the rat ate rapidly at first but then more and more slowly as time passed" (Skinner, 1979, p. 59). Skinner showed his data to W. J. Crozier, of the Harvard Department of Biology. He later wrote to his parents about Crozier's response to his discovery.

> Crozier is quite worked up about it. It is a complicated business and deep in mathematics. In a word, I have demonstrated that the rate in which a rat eats food, over a period of two hours, is a square function of the time. In other words, what heretofore was supposed to be "free" behavior on the part of the rat is now shown to be just as much subject to natural laws as, for example, the rate of his pulse. (Skinner, 1979, p. 59)

In retrospect, this may seem a straightforward result, but it is one unlikely to have been achieved outside an experimental program. If Skinner had simply watched the rat, he would have seen it do a variety of things. Besides opening the door and eating pellets, it moved around the box, sniffed in the corners, pressed its forepaws against the wall, stopped occasionally to drink water or gnaw on the water trough, and so on. Every once in a while, however, it went to the door, opened it, and consumed a pellet. Then it decided to do something else for a while. What Skinner's apparatus recorded over the course of two hours however was a surprising regularity to the rat's acts of opening the door and consuming pellets. In the midst of all its other activities, the rat managed to open the door at just the right time to create a smooth curve on the cumulative record.

Skinner concluded that the probability of the rat's performing the door-opening response during that two hour period oscillated with wavelike regularity from one to zero. Although this regular fluctuation in the probability of the response is a function of an environmental variable (the changing degree of food deprivation), common sense is unaware of any such relationship. People are not naturally adept at discriminating the relevant relationships. In a natural setting, the variables controlling the behavior do not necessarily change at a revealing rate. The food deprivation level of Skinner's rat was at a high level at the beginning of that first operant experiment, but gradually decreased in an orderly fashion--because Skinner gave the rat only one small pellet of food for each act of opening the door. If he had simply delivered a day's ration of pellets for a single response, the quantitative connection between the controlling variable (level of deprivation) and behavior (opening the door) would have been overlooked. Unaided observations would not have been likely to note the quantitative relationship between the two, even in this controlled setting. Only the use of the cumulative recorder made the relationship visible.

Skinner's dispute with common sense is asymmetrical. Skinner can give a rigorous account of a certain narrow slice of experimentally arranged behavior about which common sense has nothing rigorous to say. Common sense on the other hand has an interpretation of virtually all operant behavior, but no experimentally established quantitative account. Indeed, the particular version of common sense that Skinner is targeting could produce such an account only upon pain of inconsistency: it posits an inner agent precisely because no causal account is thought to be possible. There is no physical variable of which such behavior is a function. And if common sense can usually supply a plausible set of beliefs and desires to explain whatever someone does, these beliefs and desires do not actually compel the agent to act. If there is a causal relation in common sense explanations, it is not from reasons to actions, but from actions to reasons. A free agent can decide what beliefs to hold, what goals to adopt, and what reasons to act upon. And if these beliefs, goals, and reasons constrain the agent's actions (and in this sense explain them), this is only because of the agent's free choices. Ultimate responsibility for these choices however rests with the agent, whom common sense views as capable of initiating causal chains through an act of will. Even an agent's mind does not determine his actions, for a free agent can make up his own mind. This is the implicit concept of the inner agent that Skinner is at pains to refute at the very beginning of his experimental program.

Let us pause to summarize what Skinner accomplished vis-a-vis common sense. First, he joined issue with its distinctive (libertarian) conception of behavior. Second, and more importantly, he neutralized the sources of empirical support for that conception. Common sense is based on ordinary experience. That experience comes in two forms: subjective and objective. Skinner dismisses subjective experience on grounds it is unreliable (cf. the argument about acquisition of psychological terms). He neutralizes objective experience as well by showing it to be unrevealing: ordinary settings simply do not control enough variables and common sense simply does not keep track of changes over a long enough period of time to resolve any of the issues under dispute. Only carefully constructed experiment will do. And radical behaviorism may reasonably hope to hold its own against common sense when the issue is decided on these grounds.

Indeed, as the program of research developed, more and more detailed relationships were uncovered between the environment and operant behavior. These relationships add layer upon layer of argument for the contention that voluntary behavior is similar to other natural phenomenon: it has natural causes that are themselves part of the system of causes and effects studied by natural scientists. Hence, there is no need to posit an uncaused cause, a free inner agent, whose decisions are beyond the scope of scientific analysis.

This is the first philosophical conclusion drawn by Skinner from his empirical research. The conclusion is consistent with radical behaviorism, but is by no means equivalent to it. Mentalism comes in at least two versions: a free agent version and a deterministic version. The fact that behavior analysis continues to discover more and more causal regularities between the environment and operant behavior tends to disprove the free agent version of mentalism. This is an inductive (non-deductive, ampliative) argument that gathers strength as the program progresses.

There is nothing especially unique to Skinner's argument in this regard. Many scientists from a variety of disciplines have put forward evidence that physical variables control voluntary behavior. Few, if any, natural scientists are going to accept common sense's anecdotal evidence for the existence of an autonomous agent as outweighing the steadily accumulating evidence to the contrary from the biological, psychological, and social sciences. Thus, insofar as Skinner is addressing this aspect of his argument to scientists, he is simply adding his own pebble to the mound of evidence scientists have produced for naturalism. And insofar as he is addressing it to non-scientists, he argues on behalf of the entire scientific community, and not simply on behalf of radical behaviorists.

II

The aspect of Skinner's radical behaviorism that is controversial among scientists has to do with his explanation of *how* natural causes exert control over voluntary behavior. Most cognitive psychologists would agree with Skinner that the environment plays an important causal role in bringing about voluntary behavior, but they would insist that a full account of that role requires reference to mental states and processes. These cognitive psychologists are just as experimental and deterministic as Skinner. Like Skinner they reject the existence of autonomous man. But unlike Skinner, they see no way to give a full account of voluntary behavior without mental states and processes.

Joining Issue with Cognitive Psychology. The question is whether mental entities are necessary components in the explanation of behavioral regularities themselves. That is, even supposing that there are valid environment-to-behavior regularities, one can still ask how to account for these regularities. And this is where the scientific case for mentalism is strongest. Schedules of reinforcement, for example, create a high degree of order in behavior--perhaps higher than has been produced by any other means (Staddon, 1967, 1973). Certain schedule effects are so vivid and reliable that Murray Sidman (1960) suggested that a psychologist could use them to test equipment. Just program the equipment for one of the basic schedules, expose an animal to it, and check the results. If it generates the expected cumulative record, everything is in working order. Otherwise, there is probably something wrong with the equipment!

The question thus arises of how schedules generate such predictable effects. When a complex cause produces a complex effect, it is good scientific practice to analyze the causal process into an interaction of simpler processes (Skinner, 1953b). Skinner knew that the leading candidate for such an analysis was the hypothesis that schedules produce their effects through a system of underlying cognitive processes. The point of his most ambitious experimental work, *Schedules of Reinforcement* (Ferster & Skinner, 1957), was precisely to argue that schedule effects could instead be accounted for on the basis of the complex interaction of elementary behavioral processes. The philosophical purpose of this hypothesis was to defend radical behaviorism. If schedule effects could be accounted for on the basis of elementary behavioral processes, then inasmuch as elementary behavioral processes are plausible candidates for physiological (as opposed to psychological) analysis, there would be no remaining role for cognitive processes in the overall analysis. This was Skinner's most powerful argument on behalf of the thesis that mental states do not exist. The argument appears in Ferster and Skinner (1957), and is the outcome of a sustained experimental investigation that Skinner calls his "Golden Age as a behavioral scientist" (1984a, p. 133).

There actually was a second type of behavioral regularity that Skinner targeted, and this was almost the epistemological inverse of schedule effects. Instead of being a little known aspect of animal behavior, it is a well known aspect of human behavior. And instead of being in the domain of controlled experimentation, it is in the domain of common sense. I am speaking here of the regularities associated with verbal behavior. These regularities are similar to schedule effects in the important respect that they are complex. Skinner worked on a behavioral account of verbal behavior ever since his dramatic interchange with Alfred North Whitehead during Skinner's graduate school days. After setting the problem aside several times, he finished his account of verbal behavior in 1955 and published it in 1957. In it he argues that although the regularities of verbal behavior appear at first to require mentalistic processes to account for them, a closer look reveals they can be accounted for behaviorally.

The year 1957 thus constitutes a kind of high water mark for the advancing tide of radical behaviorism. In that year Skinner published his two most ambitious accounts of complex behavioral phenomena. These accounts not only confront mentalism's challenge to explain complex behavior, but they are theoretical in a sense that little else by Skinner before or since has been. In describing his collaboration with

Ferster on schedule effects, he notes that part of the time they worked as "Baconians," but at other times as "Galileans" (1984a, p. 73). This is to say, they did not just systematically explore a variety of schedules to see how animals would react to them (as a Baconian would), but they also tested an hypothesis about why these schedules had the effect they did (as a Galilean would). Their hypothesis was not inferred on the basis of a logic of discovery, it was invented. And it did not consist of new experimental regularities (such as the fixed-interval scallop or the fixed-ratio stair step), but consisted of an hypothesis that attempted to explain such regularities. The work on verbal behavior was theoretical in a similar sense. It consisted of an explanation of certain regularities of verbal behavior. Although these regularities were not discovered experimentally or stated rigorously, his explanatory account of them was virtually identical with the Ferster/Skinner account of schedule effects. He took elementary behavioral processes that had been discovered in the animal laboratory, and used them to account for complex behavioral regularities. The empirical case against mentalism had never advanced so far on so wide a front, and perhaps never will again.

Explaining Schedule Effects. The cumulative record of an organism's operant behavior on a certain schedule of reinforcement summarizes the effects of that schedule upon the frequency of operant responding.[1] Fixed-ratio schedules cause one pattern of responding, fixed-interval cause another, and so on. Material for this type of causal account is abundantly supplied in Ferster and Skinner's (1957) survey of schedules. Skinner elsewhere imagines a complete catalog of operant behavior that would specify all possible schedules and the cumulative record generated by each. (Actually, there are an infinite number of schedules, so such a catalog would be impossible to complete.) He is not content, however, to stop here.

---

[1]I am ignoring a lot of detail here. For example, the inverted scallop appears only at certain intervals and only in the absence of discriminative stimuli in the environment that mark the end of the interval (see Ferster & Perrott, 1968).

> A thorough analysis must go further. *Why* does a given schedule yield a given performance? . . . We need to examine the way in which a particular schedule actually affects the organism. (Skinner, 1953b, quoted in Ferster & Perrott, 1968, p. 343.)

This type of question requires a quite different type of account that specifies a process producing the causal regularity in question.

Ferster and Skinner (1957) were aware that such a question requires a theory to answer it. What distinguishes their theory, however, is that it does not appeal to underlying mental processes.

> A more general analysis . . . which answers the question of *why* a given schedule generates a given performance . . . is in one sense a theoretical analysis; but it is not theoretical in the sense of speculating about corresponding events in some other universe of discourse. It simply reduces a large number of performances generated by a large number of schedules to a formulation in terms of certain common features. It does this by a closer analysis of the actual contingencies of reinforcement prevailing under any given schedule.

Ferster and Skinner summarize the process they think can account for schedule effects as follows:

> A schedule of reinforcement is represented by a certain arrangement of timers, counters, and relay circuits. The only contact between this system and the organism occurs at the moment of reinforcement. We can specify the stimuli then present in purely physical terms. These must include a description of the recent behavior of the organism itself. The extent to which features of the present or immediately past environment actually enter into the control of behavior is an experimental question. Under a given schedule of reinforcement, it can be shown that at the moment of reinforcement a given set of stimuli will usually prevail. A schedule is simply a convenient way of arranging this. Reinforcement occurs in the presence of such stimuli, and the future behavior of the organism is in part controlled by them or by similar stimuli according to a well-established principle of operant discrimination. . . . The behavior of the organism under any schedule is expressed as a function of the conditions prevailing under the schedule, including the behavior of the organism itself. Some schedules lead to steady states, in which repeated reinforcements merely emphasize the control being exerted by

> current conditions. Under other schedules, reinforcement under one set of conditions generates a change in performance leading to a new condition at the time of reinforcement. The result may be a progressive change or an oscillation. . . . The primary purpose of the present book is to present a series of experiments designed to evaluate the extent to which the organism's own behavior enters into the determination of its subsequent behavior. From a foundation of such results we should be able to predict the effect of any schedule. (Ferster & Skinner, 1957, 2-3)

In brief, Ferster and Skinner think they can explain schedule effects on the basis of elementary behavioral processes (viz., reinforcement, extinction, stimulus discrimination, conditioned reinforcement, and differentiation of form of responses--see pp. 8-10), in combination with a careful specification of the stimulus (in some cases, inferred rather than observed) at the moment of reinforcement.

This hypothesis provides an empirical foundation for a particularly strong version of radical behaviorism. By implying that we can account for schedule effects without appealing to underlying mental processes, it directly challenges cognitive psychology in a rigorous experimental context.

Consider the characteristic inverted scallop pattern of fixed-interval schedules. Ferster and Skinner account for this pattern by assuming that the pigeon is capable of discriminating intervals of time. Because responses occurring at the beginning of the interval never deliver reinforcement, such responses have a low rate. Because the first response to occur after the interval expires delivers reinforcement, responding towards the end of the interval has a high rate. The gradual acceleration in responding from the beginning of the interval to the end is the result of the pigeon's imperfect ability to form a temporal discrimination between the beginning of the interval the end of the interval. As the duration since the last reinforcement increases, the temporal stimulus becomes increasingly similar to the stimulus that accompanied previously reinforced responses. The gradual increase in the rate of responding can then be explained as an instance of the elementary behavioral process of stimulus induction along a generalization gradient.

This process had been studied in simpler contexts. If for example a tone with a certain frequency is present when (and only when) operant responding will deliver reinforcement, then the rate of responding in the presence of a tone having a different frequency will depend upon its similarity to the original tone--the greater the similarity, the higher the rate of responding. Significantly, as one gets closer and closer to the frequency of the original tone, a given increment in cycles-per-second has an increasing effect upon the rate of responding, so that a steady change in pitch results in an accelerating increase in the rate of responding. Ferster and Skinner suggest that the same behavioral regularity is at work in a fixed-interval schedule. As the time since the last reinforcement increases, a stimulus correlated with the passage of time functions in a manner similar to a gradual increase in pitch along a generalization gradient. This hypothesis explains not only an increased rate of responding as one moves into the interval, but also the steady acceleration of that rate--i.e., it explains the characteristic scallop in the cumulative record.[2]

Ferster and Skinner are not sure what the stimulus that correlates with the passage of time is. The simplest hypothesis is that it is some aspect of the animal's own behavior--a mediating response--that makes it possible for the animal to discriminate the approximate amount of time that has passed since the last successful response. They also, however, are willing to consider the possibility that we must posit an interoceptive stimulus that changes steadily with passage of time. Such an interoceptive stimulus would be an inferred entity. Many of the experiments in Ferster & Skinner (1957) attempt to explore the extent to which schedule effects can be accounted for without such entities. For them, this amounts to the question of the extent to which mediating behavior can be shown to form the basis of certain discriminations. But they are willing to embrace the possibility that it will be necessary to posit interoceptive stimuli to account for schedule effects.

Actually, Skinner seems to have been more willing to infer

---

[2]One could always of course ask why one gets an accelerating rate of response in conjunction with a steady increase in pitch. I am not sure Skinner ever went on record with an answer to this question, but I would speculate that his reply would have been that at about this point one passes over the line between psychology and physiology. Thus, the answer to this question would not be the responsibility of psychology.

interoceptive stimuli than Ferster was. Indeed, in an early report of some of the results of Ferster and Skinner's research, Skinner (1953b) attributes the existence of an inner clock to the pigeon--a clear example of an inferred entity. Skinner notes in his autobiography that Ferster firmly opposed the positing of such entities, and continued to search for evidence of mediating behavior long after Skinner was ready to call off the search. Thus, if Ferster & Skinner (1957) seems to favor the mediating behavior hypothesis, this does not necessarily indicate that Skinner himself was convinced schedule effects could be accounted for without inferring the existence of interoceptive stimuli.

This may seem puzzling, given the widely held opinion that Skinner thought one can do psychology without inferred entities. But actually it is only at the stage of discovering basic empirical regularities that he rules out inferred entities (Skinner, 1950). He does not object to such entities in the *analysis* of those empirical regularities, although of course he does try to avoid positing such entities if simpler alternatives are available. His radical behaviorism opposes the positing only of one particular type of inferred entity--namely, one having propositional content.

Explaining Verbal Behavior. Skinner's account of verbal behavior parallels his account of schedule effects. The main difference is that there are no regularities of verbal behavior having the same degree of empirical rigor as schedule effects. But "the basic facts to be analyzed are well known to every educated person and do not need to be substantiated statistically or experimentally at the level of rigor here attempted" (Skinner, 1957, p. 11). Since rigorous experimental data are lacking, it is not possible to prove that verbal behavior is due to certain causes. What can be done is to interpret it in a manner consistent with the concepts and processes that have proven useful in an experimental analysis of animal behavior. Such an interpretation is similar in its relation to scientific knowledge of elementary processes to that which a physicist might give of the thermodynamics of pouring cold cream into a cup of hot coffee at the breakfast table (Skinner, 1953b). A careful experimental analysis under breakfast table conditions may be out of the question, but the basic regularities are well known, and the point of an interpretation is simply to synthesize the complex phenomena of ordinary life in terms of elementary processes observed in controlled experimental settings. Prediction and control are not the issue, but rather the ability of the basic science to offer a comprehensive account of familiar phenomena.

A given episode of verbal behavior can be compared to a pigeon's performance on a schedule. The performance occupies a sizable stretch of time. It is complex, consisting of a series of individual responses; these complex performances occur in certain settings; and there are regularities relating these settings to the complex responses they give rise to. In the case of schedules of reinforcement, we have such regularities as the fixed-ratio stair-step or the fixed-interval scallop. In the case of verbal behavior, the regularities are less rigorous but nonetheless well known: the English speaking guest who finds the soup lacking in flavor says "Pass the salt"; the co-worker who enters the office at the beginning of the day says "How d'you do?"; the child who sees the dog chase the cat says "Spot chased Puff"; the pet owner who sees the fur on the rug asks "Did Spot chase Puff?" These regularities play the role in Skinner's analysis of verbal behavior that is played by cumulative records in Ferster and Skinner's analysis of schedules of reinforcement.

The main thesis of *Verbal Behavior* is that the same processes that Ferster and Skinner used to account for schedule effects can also give an adequate interpretation of these aspects of verbal behavior. The difference between the processes accounting for verbal behavior and those accounting for schedule effects is supposed to be nothing but degree of complexity. People in a linguistic setting have many more responses available to them than does a pigeon in an experimental chamber. Their behavior may be subject to a larger number of reinforcers (most of which are conditioned reinforcers), and a larger number of discriminative stimuli. Nonetheless, "recent advances in the analysis of behavior permit us to approach [this complexity] with a certain optimism" (Skinner, 1957, p. 3).

> The basic processes and relations which give verbal behavior its special characteristics are now fairly well understood. Much of the experimental work responsible for this advance has been carried out on other species, but the results have proved to be surprisingly free of species restrictions. Recent work has shown that the methods

> can be extended to human behavior without serious modification. (p. 3)

The processes he is referring to are the same ones listed in the opening pages of *Schedules of Reinforcement*: reinforcement and extinction, stimulus control, conditioned reinforcement, shaping, deprivation, and a few others.

Just as an experimental analysis must begin with an attempt to define the unit of behavior, so must a non-experimental interpretation. Because verbal behavior is to be interpreted as a form of operant behavior, its basic units will be defined on the basis of behavior's effect on the environment. And just as there are many different topographies by which a lever press may be accomplished, so likewise are there many different topographies by which a given type of verbal response may be accomplished. Traditional descriptions of verbal behavior have developed taxonomies of these topographical features. Such taxonomies have their practical uses--in teaching a certain prescribed way of speaking and writing, or in assisting with translations from one language to another, for example--but they are not adequate to account for the probability of responding at a given time. An operant analysis therefore does not employ formal units. A response that might appear to be complex--such as a stock phrase or expression--may be under control of a single variable and therefore a unit. Two responses that are formally identical may be under control of different variables, and therefore members of different operants. And these units may differ from speaker to speaker. A phrase that must be composed by one speaker may exist as a unit in the repertoire of another. The basic categories of verbal behavior are thus defined functionally. The three most important of these categories are the mand, the tact, and the autoclitic.

Skinner defines the *mand* as "a verbal operant in which the response is reinforced by a characteristic consequence and is therefore under the causal control of relevant conditions of deprivation or aversive stimulation" (pp. 35-36). The utterance "More soup" by a hungry diner is likely to be a mand under the control of the speaker's hunger (although not if it is in the script of a play or a ruse to get the cook to leave the room). In general, a mand is a response which has been reinforced by a certain characteristic consequence, which consequence it has come to "specify" (p. 83).[3] The *tact* is "a verbal operant in which a response of a given form is evoked (or at least strengthened) by a particular object or event or property of an object or event" (pp. 81-82). This is a relationship of stimulus control. A certain aspect of the environment has come to exercise partial control over the probability of a certain type of response as a result of being correlated with reinforcement. The type of reinforcement however is variable, whereas in the case of the mand it is constant. Skinner cautions that this is not the same relationship as "refers to" or "denotes." For example, the standard greeting "How d'you do?" may function as a tact, although it does not refer to or denote anything.

> Roughly speaking, the mand permits the listener to infer something about the condition of the speaker regardless of the external circumstances, while the tact permits him to infer something about the circumstances regardless of the condition of the speaker. (p. 83)

Some of the usages to which the term tact applies would be described by the linguist or logician as referring to or denoting something. But Skinner emphasizes that he is "interested in finding terms, not to take traditional places, but to deal with a traditional subject matter" (p. 115). So tacting is not a synonym for referring, nor is the latter a subcategory of the former.

A single form of response may function sometimes as a mand and other times as a tact. A child's utterance of "doll" may upon one occasion be a mand to retrieve a lost toy, and upon another be a tact occasioned by "What is this?". Since these are different units of behavior, they are acquired separately, and if we observe the child emitting the mand we should not expect her spontaneously to possess a "corresponding tact of similar form" (p. 187).

The third basic functional category of verbal behavior is the autoclitic. Technically, the autoclitic is a sub-category of the tact, in which "we tact our own verbal behavior, including its functional [i.e., causal] relationships" (p. 314, interpolation mine). When behavior is autoclitic, part of the behavior of the speaker functions as a variable

---

[3]Presumably it "specifies" this consequence by tacting it.

controlling another part. We have already seen that behavior of this sort played a central role in the account of schedule effects. And it plays a similar role in the account of verbal behavior. In particular, the complex behavior that falls under the heading of composing novel grammatical utterances is autoclitic.

> There are at least two systems of responses, one based upon the other. The upper level can only be understood in terms of its relations to the lower. . . . [Complex compositional behavior can] be analyzed in terms of behavior which is evoked by or acts upon other behavior of the speaker. (Skinner, 1957, p. 313)

On Skinner's account, composition is a two-step process. First the speaker produces one or more fragmentary responses. These responses may be tacts and/or mands (or one of the other categories of verbal behavior--echoic, textual, intraverbal), but they are not emitted "until they have been dealt with autoclitically" (p. 346). Aspects of sentence construction such as word order, inflection, subordinate clauses, and so on, are the result of autoclitic processes which operate upon the "primordial responses" supplied by the first step.

> Suppose a speaker is primarily concerned with the "fact" that "Sam rented a leaky boat." The "raw" responses are *rent*, *boat*, *leak*, and *Sam*. The important relations may be carried in broken English by autoclitic ordering and grouping: *Sam rent boat--boat leak*. If we add the tag *-ed* to *rent* and *leak*, as a minimal tact indicating "past time," and the articles *a* and *the* to serve a subtle function in qualifying *boat*--in answer, say, to the anticipated query, *What boat?*--we get: *Sam* rented a boat. The boat leaked. Other manipulative autoclitics, including punctuation, produce at least seven other versions. (p. 347)

Skinner offers this as an example of the process by which novel sentences are constructed.

Analyzing the subject's own behavior as a discriminative stimulus controlling the emission of other behavior is a standard behaviorist tactic for dispensing with mental entities. Skinner's use of it breaks new ground, however, by using behavior that has not yet been emitted at the time it functions as a stimulus. In fact, it is not only the soon-to-be-emitted responses that can serve as discriminative stimuli, but also the speaker's own functional relationships with the environment. The autoclitic aspect of the complex response is thus sometimes under the control of the functional relationships that control other aspects of the response. The non-autoclitic components of the response are the primordial behavior upon which the autoclitic operates. As a result of the autoclitic process, the behavior itself emerges in a different form, and the autoclitic aspects of the total response are reinforced as a result of their ability to alter the effect of the primordial responses on an audience.

Skinner insists that "the possibility that we may tact our own verbal behavior, including its functional relationships, calls for no special treatment" (p. 314), but one may wonder why he does not wait for the behavior actually to occur before ascribing discriminative control to it. That is how he analyzes the schedule behavior of pigeons. Why not follow the same pattern for the verbal behavior of humans? The answer, it seems, is that Skinner is steering around a problem pointed out by Lashley (1951) in a classic paper on serial order in behavior. Lashley had shown that any analysis of grammatical ordering which assumed a left-to-right process of intraverbal chaining was inadequate. Sequential ordering of verbal behavior could not simply be based upon the process of taking the first part of a sentence as a discriminative stimulus controlling the construction of later parts. In deference to this demonstration, Skinner "puts the necessary controlling variables in the interrelationships among the fragmentary 'primary' verbal responses which are simultaneously, not serially, available to the speaker" (MacCorquodale, 1970, p. 95). He thereby purports to account for the complex, multiply embedded grammatical dependencies that Chomsky would soon thereafter cite as conclusive evidence of the inadequacy of a purely behavioral account of grammatical ability (i.e., of any account that could be modeled by a left-to-right finite state Markov process).

Skinner's theory is so far removed from the usual behavioral approach that reviewers saw immediately he was admitting the need to posit inferred states and processes. Charles Osgood (1958) for example wrote an appreciative review of the book that welcomed Skinner to the camp of those who found it necessary to posit nonverbal mediational processes in order to account for complex verbal behavior. Charles Morris (1958) found that "the sharp distinction between Skinner's approach and those who stress the role of intra-organismic processes tends to break down" in Skinner's discussion of verbal behavior (p. 214). And even though Chomsky sometimes treats behavioristic theory as equivalent to a finite state grammar (e.g., Chomsky, 1979), Miller and Chomsky (1963) formalized Skinner's theory as incorporating a context-free grammar, thereby implying that it makes use of relatively sophisticated syntactic processes. E. F. Segal (1977) even suggests that Skinner's autoclitic relations are equivalent to Chomsky's transformations, and Skinner (1980) asks in his *Notebooks* whether Chomsky's deep structures are not simply primordial verbal behavior before autoclitics are added.

The truth of the matter seems to be that Skinner's theory of verbal behavior has pushed its way into uncharted territory between behavior theory and cognitive theory. I doubt whether we have the conceptual tools to say whether the theory is cognitive or behavioral. In any event, the main burden of defending radical behaviorism during this classical era always rested on the analysis of schedule effects, and nothing in our discussion of verbal behavior indicates that it could have been otherwise.

III

In a special issue of the *Journal of the Experimental Analysis of Behavior* published in 1984, members of the field summarized recent trends. What they portrayed was a program of research that had moved rapidly on many fronts, but had stalled on precisely those projects linked to radical behaviorism. Jack Michael, for example, notes that the interpretation of verbal behavior outlined in Skinner (1957) has never inspired a significant body of research. Most of the work by behavior analysts on verbal behavior "could easily have been conceived without the benefit of the distinctions Skinner makes" (Michael, 1984, p. 369). And the most interesting line of research actually turns the strategy of *Verbal Behavior* on its head, and takes verbal behavior as an independent variable for explaining the effect of schedules, rather than explaining verbal behavior as a special case of schedule effects (Matthews, Shimoff, Catania, & Sagvolden, 1977;

Catania, Matthews, & Shimoff, 1982). Jackson Marr (1984) writes that despite the progress of behavior analysis on many fronts, the grand project outlined by Ferster and Skinner is as distant from completion as ever.

> In fact, the problem of isolating the controlling variables of schedule performance has been of immense difficulty. It has turned out that the behavior engendered by easily described schedules is not so easily analyzed. (p. 358)

As a result, "some researchers have suggested that we abandon this schedule-analysis effort altogether" (p. 358).

Indeed Michael Zeiler thinks this effort has already been abandoned. He opens his chapter on schedule research by referring to them as "the most powerful independent variables ever seen in psychology" (Zeiler, 1984, p. 485). He then notes some recent discoveries. Schedules can alter the effects of other variables in unexpected ways. Administration of a given drug can increase response rate on one schedule and decrease it on another. Electric shock normally decreases responding upon which it is dependent, but will maintain responding and produce a scalloped record when delivered on a fixed-interval basis. Schedules themselves can become components of schedules, with the higher-order schedules showing the usual properties of their type. These are important unanticipated findings.

There is, however, one major disappointment in schedule research. No progress has been made on the analysis of schedule effects that was pioneered by Ferster and Skinner (1957).

> So many experiments are relevant to these various efforts that they cannot be discussed here. Suffice it to say that we still lack a coherent explanation of why any particular schedule has its specific effects on behavior. . . . Whether the explanation has been based on interresponse time, reinforcement, reinforcer frequency, relations between previous and current output, direct or indirect effects, or whatever, no coherent and adequate theoretical account has emerged. Forty years of research has shown that a number of variables must be involved--schedule performances must be multiply-determined--but they provide at best a sketchy picture and no clue as to interactive processes. (Zeiler, 1984, p. 489)

As if this lack of progress were not enough, there is the further complication introduced by the discovery that schedules themselves can alter the effects of certain component variables (as illustrated by the apparent ability of fixed-interval schedules to cause electric shocks to function as reinforcers). This raises the specter of an infinite regress, for "each presumed variable can itself only be studied in the context of a schedule that presumably would have to be analyzed itself!" (p. 490).

Zeiler concludes that perhaps the attempt to give a behavioral account of schedule effects was a mistake.

> A given schedule has such uniform and predictable results that laws of schedules can be stated. This is no mean contribution for a science in which such precision is unparalleled. Attempts to explain *why* schedules have their effects in terms of still lower-order functional relations make no scientific sense. Unless new classes of fundamental events can be discovered (witness what DNA did for our understanding of genetic mechanisms), the more promising perspective is to try to formulate more abstract integrating principles. . . . As of now, schedule research, at least in a scientifically interesting form, is moribund. To all appearances, schedules are used as tools to study "more interesting" problems, but in and of themselves are of little apparent interest. . . . We have, for example, The Law of Fixed-Interval Schedules and The Law of Fixed-Ratio Schedules, but we will not be able to analyze these laws at a more molecular level. (pp. 490-491)

"The role of theory," he suggests, might be "to integrate these laws at a higher level," as exemplified by Herrnstein's (1970) matching law (Zeiler, 1984, p. 491). Or perhaps another type of theory, one referring to underlying processes, will be able to tackle the problem of schedule effects. If so, behavior analysts should be ready to embrace such a theory. For "the experimental analysis of behavior entails methodological commitments involving the detailed study of individual organisms and what constitutes good data; it involves no necessary commitments as to what kind of theory is appropriate" (p. 492).

The behavior analysts who maintain an interest in analyzing schedule effects are those who take an interest in underlying states and processes. Ben A. Williams (1984), for example, believes it would be useful to refer to underlying states in explaining how reinforcement affects behavior.

> Recent research has shown clearly that some consideration is required of the "state" of the organism at the moment that learning experiences occur, because very different outcomes may result from identical procedures depending upon the subject's previous history. Speculations about the internal processes mediating such history effects have led to important recent advances and should be a major component of future accounts of stimulus control. (p. 482).

In other words, Williams is proposing that the best available solution to the problem of analyzing schedule effects is a cognitive one. Charles P. Shimp (1984) agrees, concluding that the time has come for "increasing ties with cognitive psychology through appeals to cognitive mechanisms" (p. 418).

Not everyone draws the conclusion that it is time to embrace mentalism. Kennon Lattal and Peter Harzem (1984), the editors of the special issue, explicitly note their disagreement with Shimp and Williams. But no one disputes the end of the grand tradition of schedule research. The consensus over this development is by now so entrenched that it is easy to overlook the significance of it. What it suggests is that radical behaviorism, in so far as it prescribes specific research priorities, has reached a dead end. This goes far beyond the typical situation in which a scientific theory faces difficult unsolved problems. It suggests that the problems in this case are so intractable that it is time to quit working on them. The costs so outweigh the benefits that researchers do well to turn their attention elsewhere. There is no specific empirical result, no crucial experiment, that refutes Skinner's grand hypothesis. Instead, a generation's worth of research has led to no positive results and many negative results. The effect on behavior analysts has been a sense of profound frustration.

If there were no viable alternatives, perhaps work on the Ferster/Skinner hypothesis would continue anyhow. But there are alternatives. One is to keep asking the same question, but to turn to cognitive theory for help in answering it. This appears to be the course suggested by Williams and Shimp. The other, and by far more popular, alternative is to shift the focus of research. Instead of analyzing schedules into component parts in an attempt to explain them, use them as independent variables that can be synthesized into larger wholes or can interact with other variables.

Either way, the implications for radical behaviorism are clear. It can no longer count on empirical support from behavior analysis. If this shakes the foundations of radical behaviorism, then so be it. Few behavior analysts are going to stake their careers on the defense of this philosophy if doing so entails devoting themselves to unproductive research while their colleagues break new ground elsewhere. So even though not all behavior analysts have abandoned radical behaviorism, the defense of this philosophy is no longer intimately tied to the behavior analytic program of research.

[illegible]

R.J. Nelson (1969 [illegible] 1984, [illegible]) [illegible] isomorphism is the basis [illegible] mental states but claims that [illegible] equivalent to [illegible] am sympathetic to Nelson's desire to show there is a certain kind of compatibility between mental and behavioral theories. His own treatment [illegible] and subtlety that I cannot do it justice here. I am not, however, inclined to believe that the isomorphism in question extends to all cognitive states of interest to mentalistic psychology, i.e., I suspect that Nelson has overgeneralized.

Consider, for example, Fodor's mentalistic coke machine. We can give a behavioral definition of the readiness of the machine to dispense a coke (the output) upon the deposit of a nickel (the immediate input) by taking into account the history of past deposits [illegible] to extend this result to all [illegible] the mind can be modeled as [illegible] Turing machine, then there is an isomorphism between [illegible] environmental [illegible]

But it is questionable whether the mind can be modeled as a Turing machine. [illegible] return to Fodor's coke machine. As a model of mind, the feature of the machine that makes it so easy to turn into a Turing machine is its [illegible] desire. If the machine were a living being, it would resemble a creature that wanted only one thing, and always wanted it: namely, to dispense cokes. The problems for the program of producing behavioral equivalents of mental models only get interesting (i.e., overwhelmingly complex) when we confront a mind with complex beliefs and complex desires. What Nelson has done is show that the dispositions of a mind with complex beliefs and one simple desire can be modeled behaviorally if we permit the antecedents of our dispositional regularities to stretch out across time. He has not, however, shown this to be true for a mind with complex beliefs and desires, and I doubt whether this can in fact be done.

In general (i.e., except for simple organisms in simple environments), behavioral regularities and mental regularities are non-equivalent. The relationship between the [illegible]

CHAPTER TEN

THE ANALOGY WITH NATURAL SELECTION

Skinner's first published reference to an analogy between natural selection and operant conditioning occurs in *Science and Human Behavior* (Skinner, 1953a), where he distinguishes between innate behavior which arises at the level of the species, and operant behavior which arises at the level of the individual organism.

> In both operant conditioning and the evolutionary selection of behavioral characteristics, consequences alter future probability. Reflexes and other innate patterns of behavior evolve because they increase the chances of survival of the *species*. Operants grow strong because they are followed by important consequences in the life of the *individual*. Both processes raise the question of purpose for the same reason, and in both the appeal to a final cause may be rejected in the same way. A spider does not possess the elaborate behavioral repertoire with which it constructs a web because that web will enable it to capture the food it needs to survive. It possesses this behavior because similar behavior on the part of spiders in the past has enabled *them* to capture the food *they* needed to survive. A series of events have been relevant to the behavior of web-making in its earlier evolutionary history. We are wrong in saying that we observe the "purpose" of the web when we observe similar events in the life of the individual. (p. 90)

The analogy shows how operant psychology can purport to explain purposive behavior without referring to the goals and intentions of an agent. Like natural selection, operant conditioning can account for design without a designer. Instead of positing control by a pre-existing design, both theories explain adaptations on the basis of prior environmental consequences. Natural selection then stands to the phylogeny of purpose as operant conditioning stands to the ontogeny of it.

Skinner (1953a) draws the analogy once again several hundred pages later.

> We have seen that in certain respects operant reinforcement resembles the natural selection of evolutionary theory. Just as genetic characteristics which arise as mutations are selected or discarded by their consequences, so novel forms of behavior are selected or discarded through reinforcement. (p. 430)

This passage explicitly states what had been implicit in the preceding discussion. Both natural selection and operant conditioning are the result of variation and selection. Organisms and (operant) responses get replicated in forms that resemble, but are not completely like, earlier instances. A certain amount of variation thus constantly occurs. Variations with better environmental consequences are likely to have higher rates of reproduction or reinforcement, and therefore are likely to occur more frequently in the future. These are marginally better equipped to be effective within their environment. After many cycles of such marginal improvements the result may be so unique and yet so well adapted to its environment that it appears to be the result of an intelligent force. Actually, however, it is due to nothing but the blind process of variation and selection. Thus, operant conditioning allegedly stands to the ontogeny of novel adaptive responses as natural selection stands to the phylogeny of novel adaptive traits. Selection by consequences eliminates the need for a creative, intelligent force to explain either one.

I

This seems to be all that Skinner has to say about the analogy until the mid-60's, when he begins increasingly to rely upon it to support radical behaviorism. He notes, for example, that the problem of explaining purposive behavior invites us to infer that the organism behaves as it does because "it intends to achieve, or expects to have, a given effect; or its behavior is characterized as possessing utility to the extent that it maximizes or minimizes certain effects" (Skinner, 1963b, p. 105).

Thus, we often find ourselves explaining behavior by reference to an inner surrogate of the typical consequences of past responses. This surrogate is the prototypical mental entity. It has a content which represents the organism's past experience. If pecking the key has been followed by the delivery of food, the organism forms an inner representation of this contingency. This representation functions as the belief or expectation that pecking the key will deliver food. It is in the animal and available to serve as the cause of pecking when at some later time the animal gets hungry. The analogy with natural selection, however, helps us conceive of the possibility that there is a way to account for the utility of such behavior without appealing to inner representations.

Skinner furthermore asserts that selection by consequences dispenses with the need to posit a responsible agent, whether it be God at the level of phylogeny or inner self at the level of ontogeny. The latter has important practical implications. If the environment is responsible for our behavior, there is little justification for holding people responsible for their actions or urging people to take responsibility for solving their problems. Such practices are simply weak forms of behavior modification. Although sometimes effective, they fail precisely when we need them the most--in the difficult cases. The compulsive gambler, the drug addict, the alcoholic, the abusive spouse, the unmotivated student are beyond the reach of these hortatory methods of control associated with concepts of freedom and responsibility (Skinner, 1971). Acceptance of the selectionist view is thus supposed to imply a profound shift in how we approach social problems.

Such amplifications of the analogy have continued to occur (see especially Skinner, 1981), so that by 1984, when *The Behavioral and Brain Sciences* published a special issue on the "Canonical papers of B. F. Skinner," the analogy with natural selection had become Skinner's principal source of support for radical behaviorism. This is especially clear in his replies to the numerous peer commentaries.[1]

---

[1]Skinner's radical behaviorism is sometimes interpreted to mean that no appeal to underlying mechanisms is necessary to explain behavior, but if so, this position is not necessarily supported by the analogy with natural selection. The assumption that natural selection dispenses with the need to refer to underlying mechanisms or

By decisively severing the alleged link between radical behaviorism and logical positivism, and by simultaneously forging a link with recent developments in the philosophy of biology--a field rapidly gaining influence within the philosophy of science--the analogy wins serious consideration for radical behaviorism from people who otherwise would be uninterested. This is not simply because the technique of explaining design without a designer has been worked out in detail by biologists. It is also because there is heightened awareness that selectionist theory has unexpected subtleties. The theory of natural selection was eclipsed for an extended period in the late 19th and early 20th centuries, during which time biologists thought it too weak to account for the twists and turns of evolution (Bowler, 1983). But the theory returned in the

---

proximate causes is at best a contentious one (see Sober 1983; Kitcher, 1989). For there are other forces and constraints besides selection that influence evolution, and some of these cannot fruitfully be discussed without talking about or examining underlying mechanisms or processes. Meiotic drive and differential rates of mutation are clear examples of underlying forces that fit this description; pleiotropy and heterosis are clear examples of underlying constraints that fit.

Perhaps some will counter that we nonetheless could give what is by and large an accurate account of evolution in terms of selection alone (cf. Sober's 1987 definition of adaptationism). If so, then despite the existence of other forces and mechanisms, selection would remain the dominant factor determining the direction and rate of evolution. While this may be true, it does not imply that there is never a need to refer to underlying processes or mechanisms. The Darwinian view implies only that for certain purposes, we may ignore all forces but selection. But if one wants to give a complete account of the independent variables which influence the changing frequency of traits in a population, then one may need to talk about underlying mechanisms and proximate causes.

This, by the way, is a stronger statement than merely saying that eventually we may have to talk about underlying mechanisms in order to give the complete story about how the environment controls evolution--i.e., in order to explain the causal regularities defined by selectionist explanations themselves (see Cummins, 1983a, for an account of the two types of explanation involved). No one ever doubted this. The point here is a quite different one. Even some of the regularities that we want underlying mechanisms to explain are themselves about forces and constraints generated by underlying mechanisms. Hence, if a level of explanation is defined by a set of forces and constraints that interact to control a given dependent variable, then at least some underlying mechanisms and processes are part of the same level of description as the environmental forces of natural selection. The implications, by analogy, for behavioral psychology are rather obvious, and rather unbehavioral.

1930's with greater vigor than ever--a fact that provides some small comfort to radical behaviorists, who can feel the tide of history shifting against them (Catania, 1987).

More importantly for our purposes, the analogy also introduces new scope for conceptual innovation. It implies a philosophical program--a strategy for the development (and not simply the defense) of radical behaviorism. The classical version of radical behaviorism is whatever Skinner said it was. He could be wrong about the strength of the arguments supporting it, but not about its content. The analogy however provides a principled means of revising the very content of radical behaviorism, thereby making it possible for the philosophy itself to evolve.

The analogy thus has two principal uses: (1) to justify radical behaviorism, and (2) to revise it. Each use presents its own course of investigation. To justify radical behaviorism, we ask what must be the nature of operant conditioning if, by analogy with natural selection, the theory of operant conditioning is to justify radical behaviorism. This in turn raises the question of whether operant conditioning, thus conceived, is empirically defensible--i.e., whether there is scientific evidence of the occurrence of a process fitting this description. If so then radical behaviorism, as understood at the outset, receives empirical support. To revise radical behaviorism, we ask what must be the nature of radical behaviorism if, by analogy with natural selection, the process of learning as currently understood is to justify radical behaviorism. This in turn leads to the question of whether radical behaviorism thus conceived is philosophically productive--i.e., whether this revised version of radical behaviorism is capable of implying significant philosophical conclusions. Both of these methods may be incorporated into a dialectical process that goes back and forth between the two, searching for an equilibrium point at which an empirically well supported theory of learning justifies a philosophically productive conception of radical behaviorism.

II

Analogies are vague, but they do have empirical content. One of the pivotal events of modern physics--the Michelson-Morley experiment--was an attempt to test an analogy. As a source of support for radical behaviorism, the major function of the analogy with natural selection is to bear the weight previously born by the theory of schedule effects--i.e., to make the case that we do not need mental processes to account for the manner in which behavior adjusts to its changing environment. The analogy implies that an explanation can be found in a certain type of selective process. It implies that all intelligent behavior--all behavior that adapts to the changing circumstances of the organism in a manner that gives evidence of intelligence--can be explained on the basis of conditioning.

If it is to be analogous to natural selection, then conditioning must have certain key features (Amundson, 1989).

1. There must be a rich source of small, undifferentiated variations for the process of conditioning to operate upon.

2. These variations must not be directed towards adaptation, but must occur at random.[2]

3. The process of reinforcement (selection or sorting) must itself be unintelligent and nonpurposive--i.e., the determination of which response gets reinforced must be mechanical.

---

[2]As Sober (1984) notes, the requirement is not that variation literally be random, but simply that it be blind to the distinction between adaptive and non-adaptive changes. The use of the concept of randomness in this context has become so commonplace, however, that one can almost speak of it as having a special conventional meaning here.

A process of learning that meets these conditions will explain intelligent adaptation without appeal to inner representations, purposes, plans, ideas, etc.

Skinner's classical theory of operant conditioning describes such a process. Consider, for example, Skinner's (1953a) description of the process of shaping.

> Operant conditioning shapes behavior as a sculptor shapes a lump of clay. Although at some point the sculptor seems to have produced an entirely novel object, we can always follow the process back to the original undifferentiated lump, and we can make the successive stages by which we return to this condition as small as we wish. At no point does anything emerge which is very different from what preceded it. (p. 91)

This passage reaffirms Skinner's (1935b) opinion that operant behavior starts out as a form of spontaneous undifferentiated activity (condition one).[3]

The components of this activity are small grained features of behavior that Skinner (1953a) calls "elements". When a response is reinforced, it is these elements that are strengthened. These in turn interact to form new, possibly novel, responses. This relationship between past and future responses is traditionally called response generalization or response induction. Skinner knows he is not capable of solving all the problems raised by this.

> We lack adequate tools to deal with the . . . interaction among operants attributable to common atomic units. (p. 95)

What he does say about it is vague. Like most behaviorists, he thinks this relationship is based upon some kind of similarity or resemblance between past and present responses, but he does not attempt a detailed description. We may safely assume, however, that he believes the process by which novel responses emerge (whatever it may be) is unintelligent and nonpurposive (condition two).

---

[3]Richard Colker points out that a more appropriate analogy might be "as the wind and water shape the land."

Finally, there is the question of which response gets reinforced. When a reinforcing stimulus occurs, the context provides a number of possible responses that might be reinforced. Which of them gets chosen, and by what? Again, intelligence and purpose cannot be involved. Skinner's favorite explanation is temporal contiguity, according to which the response occurring just prior to the delivery of reinforcement gets reinforced. He defends this explanation from certain obvious counterexamples by positing the existence of conditioned reinforcers that bridge the observed temporal gap between response and primary reinforcer. In Skinner's original experimental chamber, for example, the mechanism that produced a pellet of food made a sound as it operated. As soon as a response that produced reinforcement occurred, the mechanism began to operate. A short time later, the food appeared. Although the pellet itself was the primary reinforcer, the stimulus that selected the response to be reinforced was the sound of the machinery. In more recent designs, a light comes on over the food hopper as soon as a successful response occurs. Skinner assumes that something similar, although more complex, happens in nonlaboratory settings. Thus, the selection of the reinforced response is unintelligent and nonpurposive (condition three). Although Skinner rarely states these conditions, he commits himself to them implicitly, if not explicitly, by his appeals to the analogy with natural selection in defending radical behaviorism. Other theorists understood these implications quite clearly, and much of the research that is cited as calling the viability of behavioral psychology into question is actually addressed to the question of whether the process of learning satisfies the preceding three conditions. Ironically, the very first experiment reported in *The Behavior of Organisms* (Skinner, 1938) is quite difficult to reconcile with the selectionist interpretation. In that experiment, Skinner releases a magazine-trained rat into the experimental chamber. The rat has already undergone adaptation to the chamber, and occasionally emits the lever-pressing response even though this response has never been reinforced by Skinner. Now for the first time Skinner connects the lever to the magazine. Each lever press will result in access to food. The animal is released into the chamber. After five minutes it presses the lever and receives a food pellet. Delivery of the pellet has no observable effect upon subsequent behavior. Over fifty minutes elapse before the rat presses the lever again. It receives a second pellet, but again without effect. Over forty-seven minutes elapse before the third lever press, and twenty-five more minutes until the fourth. Only then does the rate of lever pressing show an appreciable increase in rate. Within a few minutes the rat is pressing the lever every ten or fifteen seconds, and continues to respond at this steady rate.

This is not at all the type of behavior one would have anticipated on the basis of the metaphor of selection of responses. Why did the first three deliveries of food have no observable effect upon behavior? And why did the fourth delivery of food result in an appreciable increase in rate of lever pressing, after which it accelerates to a maximum? The metaphor of selection hardly fits what happened. Responses one, two, and three seem to have been part of a different causal process from responses four, five, six, and onward. But if we view reinforcement as a form of selection, we have difficulty understanding how this could be so.

Ferster and Perrott (1968) recognize this difficulty in their retrospective discussion of this experiment, and suggest that even with the first three responses, the delivery of reinforcement was probably selecting some aspect of behavior. These responses may have been followed by "adaptation to novel stimuli, the development of conditioned reinforcers, or the conditioning [of] successive approximations of later members of the chain such as approaching or eating from the food tray" (p. 219). Perhaps. But Skinner reports that the animal had already undergone adaptation and magazine-training before the experiment began. Thus the various parts of the chamber should have been familiar, the sound of the magazine should have been functioning as a conditioned reinforcer, and the chain of responses that culminate in eating from the food tray should have been well established. Furthermore, we have Skinner's statement that if something was being reinforced, he could not observe it. Instead, what he observed was simply a sudden change in tne rat's behavior after the fourth response. The conditioning of lever pressing, as he is at pains to point out, seems to have occurred as the result of that one response. Skinner goes on to report experiments with other rats in which it was the *first* response which resulted in conditioning, but the point is not whether it is the first or not, but that there is a sudden change in the animal's behavior. One wants to say something such as, the reinforcer contacts the response, something gets established, a connection is being made--there are many ways of describing what occurs, but talk about selection seems forced. Indeed, it positively *mis*represents what occurred by creating the false impression that the conditioning process began with the first response, whereas in fact it began only with the fourth response. Skinner himself is quite explicit about this, saying that "conditioning does not take place until the fourth reinforcement" (p. 68).

We do not need to rely upon Skinner, of course, for experimental results that seem inconsistent with the selectionist interpretation of conditioning. There is no shortage of them. Most psychologists can cite numerous experiments that purport to demonstrate that learning violates these conditions, and many behavior analysts have simply quit trying to refute such demonstrations--a fact which should not surprise us.

The conditions are of a piece with the Ferster/Skinner theory of schedule effects, and we have seen that behavior analysts have stopped working on that theory. A similar thing has happened with respect to the assumptions we have been discussing. Briefly, here are some of the reasons why.

1. **Small and undifferentiated variations.** The first major wave of scientific critiques of behavioral psychology attacked this assumption. Critics argued that animals have an innate preparedness to learn certain responses but not others, and to respond to certain stimuli but not others. Garcia, for example, found that rats that have eaten poisoned food learn to avoid food having a similar taste, even though there are many other salient aspects to the situation besides the taste of the food, and even though the painful effects associated with the poison did not occur until long after the food had been ingested (Garcia, Kimmeldorf & Hunt, 1961; Garcia & Koelling, 1966). Instead of learning responses gradually on the basis of small increments, animals learn some responses

almost immediately and other responses only after extensive training (if at all).

2. **Non-directionality of variation.** Staddon and Simmelhag (1971) were among the first behavior analysts to raise doubts about this aspect of conditioning. They suggest that, appropriate though this assumption may be in the case of evolution, its analogue in the domain of learning is probably unjustifiable. They give two reasons for this opinion. First, random variation is an inadequate basis for the process of complex problem solving. A child, for example, learns language in less time than would be possible through learning by random variation. Second, Darwinian selection is a hill-climbing process that moves in a direction that produces immediate gains at each step, but is incapable of producing long-term gains by strategically forgoing short-term benefits. In the case of learning, however, there are learned patterns of responding in which the organism forgoes small immediate gains in order to produce large delayed gains. Such patterns of responding attain global maxima that are unattainable through hill climbing processes. Intelligent animals are nonetheless capable of learning such behavior.

3. **Blind reinforcement.**[4] The leading candidate for a blind mechanism of response selection--temporal contiguity with the delivery of reinforcement--has virtually been refuted. Rescorla (1967) first showed this for classical conditioning when he demonstrated that it is the informativeness of a neutral stimulus that leads to conditioning, not simply temporal contiguity with reinforcement. Analogous results have been attained

[4]See G. C. Williams, 1966, for a discussion of the properties of the mechanism of reproduction necessary for evolution to occur.

> for operant conditioning (Hammond, 1980). This may indicate that the connection between the reinforcing stimulus and the reinforced response requires cognition.

In sum, it is very difficult to defend the three conditions empirically.

III

How is the radical behaviorist to respond to this challenge? One approach is to acknowledge that there are problems with the analogy, but to emphasize that these problems are difficult to assess. This combines well with a counteroffensive that claims processes such as shaping, stimulus control, fading, chaining, etc. can account for the phenomena mistakenly assigned to cognitive processes. This is the thesis of Skinner and Epstein's Columban Simulation Project, which attempts to show that various patterns of responding typically attributed to cognitive processes are actually under environmental control (Epstein, 1981). They believe they can demonstrate this by using standard operant techniques with pigeons to simulate complex behavior such as insight, possession of a self-concept, symbolic communication, and talking to oneself. (See Baxley & Associates, 1982, for a filmed record of this effort.) Such simulations, however, raise as many questions as they answer. At best, they plant honest doubts about the strength of classic demonstrations of cognitive processes (e.g., Kohler's insight experiments, which were performed during World War I), but they do nothing to refute carefully controlled, contemporary experiments attacking the three conditions.

A more promising approach is to question whether learning needs to be so strictly analogous to natural selection in order to support behaviorism. Staddon and Simmelhag (1971), for example, reject selection as the dominant force in ontogeny. They do not, however, reject the analogy itself. Rather, they suggest a way to amend it. Noting that evolution through selection consists of two components--variation and selection--they say that the reason why Darwin can emphasize selection is that the biological process of variation has little structure to it. An unqualified analogy between evolution and learning, they say, breaks down precisely here. For behavioral variation is the result of underlying mechanisms which tend to produce novel adaptive responses. They conclude that the theory of behavior must give more weight to principles of variation than does the theory of evolution. An adequate science of behavior will require a synthesis of these principles.

This would constitute a profound shift in the way the analogy with natural selection is used. Staddon (1983) follows through on this suggestion in impressive detail, searching for less constrained versions of the analogy that are more consistent with the known facts about learning. In this way, the analogy functions not so much as a source of empirical support for radical behaviorism as a source of ideas for how to revise it. Staddon himself abandons the term radical behaviorism, but whether we keep the term or not, the continued use of the analogy forges a link to that philosophy.

Even a classical radical behaviorist admits there must be an inner state or event that mediates the relationship between past contingencies and current responses. Natural selection acknowledges the existence of inner mediating states in the form of genes. These genes may seem superficially to resemble cognitive states, but they do not necessarily support an interpretation of inner mediating states as representations. In biology, the idea that the genome represents a preformed and predetermined entity (preformationism) has given way to the idea that development of the individual is the outcome of continuous interactions between a genetically encoded program and the environment (epigenesis). The gene for brown eyes is not a representation of brown coloration, but is simply a strand of DNA that in a certain context will give rise to brown coloration of the iris. Genes apparently encode information without being inner representations.

It is beyond the scope of this essay to pursue the ramifications of this observation for attempts to use the analogy with natural selection to interpret and revise radical behaviorism--except to note that it has a tendency to make radical behaviorism resemble pragmatism. In particular, radical behaviorism's conception of underlying states begins to acquire the same contrast with the copy theory of concepts that pragmatism's conception has. We have already noted a similarity between Skinner and Dewey, especially in their emphasis upon the active organism (Hilgard, 1956). Recently, a few radical behaviorists have taken the step of simply identifying Skinner's behaviorism with pragmatism (e.g., Schnaitter, 1984). On such an identification, it is difficult to say if radical behaviorism would continue to be the thesis that mental states (states having propositional content) do not exist, or whether it would now become a thesis about *how* mental states *acquire* their content. The justification of the latter thesis, however, would be independent of the success or failure of behavior analysis.[5]

Summary. Where does this leave the attempt to use the analogy with natural selection to interpret and assess radical behaviorism? For roughly half a century their program of research has been discovering and refining a substantial body of empirically valid environment-to-behavior principles. These principles themselves make no mention of mental states or processes, but critics have claimed that the analysis of the process underlying these patterns of behavior requires positing mental states and processes. In defense of this program's potential completeness, Skinner argued that appeal to mental processes was unnecessary for any of the legitimate purposes of psychology. His rigorous experimental program for defending this claim implied that the only underlying states or processes we need in psychology are strictly analogous to the elementary states and processes we observe in the operant chamber (Ferster & Skinner, 1957). That is to say, the only inferred entities we need are similar in kind to the entities we observe

---

[5]Some radical behaviorists take heart from the dramatic success of connectionist (parallel distributed processing) models of late. They interpret this to vindicate their anti-mentalism. But just as it is far from clear that pragmatism is a genuine alternative to mentalism, so also is it far from clear that connectionism is an alternative to mentalism. It is quite possible that connectionism may simply turn out simply to be an alternative form of mentalism. True, it does seem to reject the Chomsky/Fodor version that interprets mental content to be a representational relationship and that interprets mental processes to be a form of symbol manipulation. In this sense, the connectionist approach to mental states and processes is perhaps closer perhaps to radical behaviorism than is the Chomsky/Fodor approach--but if it is still a form of mentalism, that would seem to provide little comfort to radical behaviorists (cf., Bechtel, 1988a).

directly. At the same time, however, his speculative interpretation of verbal behavior conceded the need to infer the existence of states and processes that seem different in kind from those we directly observe.

It is difficult to avoid the conclusion that Skinner's philosophical position is either less consistent, or else less profound, than it at first seems--and perhaps both. If he thinks all behavioral phenomena can be accounted for on the basis of the elementary behavioral processes of Ferster and Skinner (1957), then how is that consistent with his theory of verbal composition? But if he thinks behavior analysis should be free to posit the primordial responses and compositional processes of Skinner (1957), how is that significantly different from the practice of cognitive psychology? If one tries to make the position consistent by discounting Skinner's theory of verbal composition, this accomplishes little so long as behavior analysts fail to make headway on the grand hypothesis that schedule effects can be explained on the basis of elementary behavioral processes. But if one takes the hint supplied by Skinner's theory of verbal composition and equates radical behaviorism with pragmatism, this does little to define or defend the mission of behavior analysis. Instead of reaching an equilibrium, the philosophy of radical behaviorism seems to be wobbling further off center.

PART FOUR

DISENTANGLING THE PROGRAM FROM RADICAL BEHAVIORISM

I feel we are machines. I couldn't be a behavioral scientist if I didn't. But not a machine like a wind-up toy. More like a leveling device on a ship. We are goal-directed machines, and this is something I think Fred doesn't understand. But then, Fred is a visionary, and visionary people are visionary partly because of the very great many things they don't see. (R. J. Herrnstein, quoted in Rice, 1968, p. 137)

The environmental causes of a given act will generally be many, largely or totally in the past, and related both to one another and to the final outcome by mechanisms of immense complexity. The attempt to bridge temporal and conceptual gaps by means of S-R explanations of the hooks-and-eyes sort, while perhaps justifiable as a working hypothesis at an early stage in the history of behaviorism, can no longer be seriously entertained. (Staddon, 1973, p. 43)

Different explanatory modes should not compete with each other, they should complement each other. (Bolles, 1984)

Radical behaviorism, in its classical form, put forward the ontological thesis that mental states do not exist,[1] then defended this thesis by attempting to demonstrate that reference to such states is superfluous to a scientific account of behavior. The extent to which this philosophy implied an agenda for behavioral psychology is evident in Skinner's research interests. At the peak of his scientific career his two major projects, Skinner (1957) and Ferster & Skinner (1957), were attempts to defend radical behaviorism by showing that complex behavioral regularities could be explained on the basis of the interaction of elementary behavioral processes. In retrospect both attempts must be judged failures. Not only have behavioral explanations of these phenomena failed to progress beyond the point where Skinner left them in 1957, but the main impact of subsequent research has been to suggest it is unlikely that complex behavioral phenomena can be explained on the basis of simpler behavioral processes of any kind.

Despite the lack of progress, however, Skinner steadfastly held to the goals of his research program. But rather than defend it on the basis of empirical results (as he did during radical behaviorism's classical era), he came to rely upon the alleged analogy between operant conditioning and natural selection. This analogy offered hope that operant conditioning could provide an environmental account of learning that is analogous to the environmental account natural selection has provided of evolution.

We have seen however that this analogy is difficult to defend. The problem is not that contingencies of reinforcement fail to exert control over behavior, but that cognition seems to play an essential role in bringing about such control--i.e., there is reason to believe that behavioral principles of the sort discovered by Skinner and refined by successive generations of behavioral psychologists cannot be explained without reference to underlying cognitive processes. Some theorists have attempted to modify the philosophy of radical behaviorism to

---

[1]As we saw above, the type of mental states said not to exist are the inferred states such as beliefs and desires that we frequently cite as explanations of behavior, and not subjective phenomena (events, actually) such as inner speech or feelings, that are objects of direct awareness.

accommodate these developments. Such modifications may render the philosophy more defensible, but they also render it less relevant to the task of explaining and maintaining the progressiveness of the behavior analytic program--a task which is central to the purpose of this essay.

The strength of operant psychology has been its ability to arrive at well defined causal regularities. It can no longer realistically claim to account for such regularities by reducing them to more basic behavioral principles. A corollary is that it can no longer realistically claim to explain away the appearance that purpose or intelligence underlies certain complex patterns of behavior. Evidently, Skinner's vision of operant psychology combining with a theory of respondent behavior (classical conditioning) and a theory of released behavior (ethology) to offer a comprehensive treatment of the entire domain of psychology is unattainable. It would leave out too much. The question thus arises of how to disentangle operant theory from radical behaviorism and to define its legitimate role within psychology as a whole. The struggle to free operant psychology from the grasp of radical behaviorism requires more than a critique of behaviorism. It also requires positive developments that carry the program in new directions and form the basis for a more adequate understanding of behavior.

This is because radical behaviorism's role has not simply been the negative one of justifying a systematic neglect of mental processes but also the positive one of setting an agenda for the program of research that would lead to further scientific progress. Thus, it is not enough simply to show that radical behaviorism, regarded as an empirical thesis, is false. Disentangling the program from the philosophy also requires positive developments that set a new agenda and forge links to practical scientific concerns. The following chapters examine such developments and chart their philosophical implications.

CHAPTER ELEVEN

TRANSCENDING BEHAVIORISM

Skinner saw himself as leading a revolutionary movement that would replace mentalistic psychology with a superior behavioral alternative. Its motto might have been, "Anything mentalists can do, behaviorists can do better." The premise underlying this motto was the thesis that mentalists are talking about something that does not exist. Skinner's scientific defense of this premise was his theory of schedule effects. His quasi-empirical defense of it was the analogy between operant conditioning and natural selection. The entire package was the philosophy of radical behaviorism.

There was a time when the typical behavior analyst was a radical behaviorist. This era is personified in the figure of C. B. Ferster, who collaborated with Skinner during his "Golden Age as a behavioral scientist" (Skinner, 1984a). It was Ferster who opposed reference to inner processes of any kind--to the point that he pressed Skinner to remove even the seemingly harmless references to inner clocks and counters from Ferster and Skinner (1957).[1] Ferster held a set of convictions that seem to have been rather widespread among behavior analysts during the early post-war period. To them, a commitment to operant psychology was equivalent to a commitment to radical behaviorism, which in turn was equivalent to a commitment to the research strategy that reached its zenith in Ferster and Skinner (1957). It is increasingly difficult to find active researchers who hold such convictions. Indeed, the ultimate goal of much discussion among

---

[1] See Skinner, 1953b, for an early report of research by Skinner and Ferster in which references to inner clocks and counters play a prominent role; see Skinner, 1984a, for an account of Ferster's opposition to such references.

operant psychologists today is to find a way to maintain the integrity of their program of research while acknowledging the legitimate discoveries of cognitive psychology.

This is a healthy development. Radical behaviorism seeks to show that we have no need to refer to cognitive states and processes in the explanation of behavior. It thereby raises issues and focuses on problems that are no longer a profitable investment of scientific resources. Evidently, there are major aspects of psychology for which a behavioral approach is inadequate. So the central thesis of behaviorism--that mental concepts have no legitimate role to play in psychology--is apparently false. The possibility exists, however, that there are aspects of psychology for which a behavioral approach is nonetheless optimal. Discovering what exactly these aspects are, however, has not been easy. To move Skinner's program forward requires not only a rejection of radical behaviorism, but a creative transformation of the concepts and theories that were associated with that philosophy.

I

The strategy of behavior analysis has sometimes seemed little more than this: do some experiments on rats and pigeons, then extrapolate the results to human beings. One might therefore assume that the potential of behavior analysis is roughly equivalent to the potential of such extrapolations.[2] The question of whether results obtained with animals will extend to human beings therefore takes on a special significance. Although radical behaviorism may not strictly speaking have assumed that such extrapolations would prove valid, the relevance of operant theory to practical human concerns gains much of its plausibility from the assumption they are. The discovery that schedule effects are consistent across species (see the end of Chapter Three

[2]This is an assumption made explicit by many critiques of operant theory, including sophisticated treatments such as Mackenzie (1977) and Schwartz & Lacey (1982). Rosenberg (1988) is not so explicit, but I find the most coherent interpretation of his critique of behavioral psychology to require this assumption.

above) instilled confidence in operant researchers that contingencies of reinforcement are ubiquitous features of behavior--including human behavior.

Thus the discovery in the 1970's that the fixed-interval scallop disappears from human behavior at about the age of four or five (Lowe, 1979) created somewhat of a crisis within operant circles. The fixed-interval scallop is perhaps the most robust of all schedule effects, so if this cannot be found in adult human behavior, one has to ask what is left of the assumption that animal results can be extrapolated to the human domain. Four or five is about the age when complex linguistic patterns emerge in a child's behavior, so the discovery supports Chomsky's (1959) claim that whatever validity may attach to operant principles in the domain of animal behavior fails utterly within the domain of normal behavior of adult human beings.

This is not, however, the only inference one might draw from the data. The assumption that one can extrapolate from animals to people has two quite different interpretations, and only one of these has drastically negative implications for behavior analysis. The assumption can be taken to mean that the same experimental procedures that are used on animals will generate the same experimental effects when they are used on human beings, or to mean that the causal principles discovered through research on animals extend to the human domain. The difference between these interpretations becomes apparent once we recall that the independent variables of behavioral principles are analogous to forces. A behavioral principle typically asserts that the amount of Y of a certain complex response is influenced by the amount of X of a certain complex stimulus. This assertion is not a universal generalization that says a certain amount of X will always be accompanied by a certain amount of Y. Instead, it states the contribution that the amount of X makes to the amount of Y. There is a significant difference between the two.

Imagine that the laws of mechanics were interpreted as universal generalizations. Then they would imply that whenever a force of a certain magnitude is applied to an object of a certain mass, the object will move. But of course, forces do not work that way. We may apply a force of the supposedly requisite magnitude to an object of the designated mass, and nothing may happen--not because the force does not have the predicted effect, but because another force is having an equal and opposite effect. In other cases, the net result may be movement, but not the movement that would have been predicted on the assumption that laws are universal generalizations. For example, we may apply a force from the west, but the object moves northeast rather than due east, because a force from the south also is at work. The same can happen with behavior. The amount of punishment following a child's acts of misbehavior may be influencing the amount of misbehavior, but there may be no visible effect because some other behavioral force (e.g., the reinforcing attention of other children) is having an equal and opposite effect. Or punishment may have an effect, but not the one we predicted, because there is an additional force at work (e.g., the availability of reinforcement outside the classroom that combines with punishment of classroom misbehavior to cause an unintended increase in truancy).

It is a truism that human behavior is more complex than animal behavior. To a cognitive psychologist this means the processes underlying behavior are more structured. To a behavior analyst, however, it means there are more forces at work. Saying there are more forces at work can itself, however, mean two quite different things. It can mean there are a larger number of operant responses, reinforcers, and discriminative stimuli involved in human behavior than one finds in animal behavior (this seems to have been Skinner's, 1953a, view). Or it can mean that there are more kinds of forces at work in human behavior than in animal behavior.

In the latter case, even well confirmed results obtained with animals in carefully controlled settings will not necessarily extend to human beings--not because human behavior is not subject to the same factors that influence animal behavior, but because it is subject to additional ones. Just because we establish a certain relationship between behavior and environment in a wide range of animal species, it does not necessarily follow that a human subject will generate the same result in the same controlled setting that an animal does. For certain other forces may impinge upon human behavior that do not affect animal behavior. Much recent conceptual innovation in the field has been motivated by the desire to incorporate such new forces into a behavioral account.

Instructional Stimuli and Rule-Governed Behavior. Human beings have, in addition to the capacities associated with reinforcers, discriminative stimuli, and operant responses, some qualitatively different capacities. Some of these capacities generate forces that interfere with the effects of classical three-term contingencies of reinforcement.

Take an obvious example. Human beings have a capacity for producing and responding to verbal stimuli. It is clear that a verbal instruction can alter human behavior. Behavior analysts have not succeeded in developing a property theory that accounts for this capacity. Presumably this capacity is in part the result of some underlying cognitive processes. Therefore, behavior analysis cannot claim to have solved the philosophical problem of intentionality, nor even the problem of how the effect of a complex sentence can be a function of the way the parts of the sentence interact. But no matter. Behavior analysts do not need a property theory of verbal behavior to study the effects of verbal stimuli.

The ability of a verbal instruction to control operant behavior can itself be treated as a functional relation subject to behavior analysis (Catania, Shimoff & Matthews, 1989). As with any functionally defined category, the instructional stimulus (by definition) makes contact with behavior. The behavior analysis does not explain how this contact comes about. Perhaps part of the story involves the application of certain heuristic strategies for sentence comprehension, plus an inductive assessment of the reliability of the speaker. Perhaps it is more complicated than that. We can set these questions aside.

Just as a given stimulus can be categorized as a reinforcer without knowing what underlying process gives it its reinforcing capacity, so also a given stimulus can be categorized as instructional without knowing what process underlies its instructive capacity. Such is the rationale for introducing the complementary concepts of instructional stimulus and rule-governed behavior. When Skinner (1969) first introduced these concepts, he treated a verbal instruction as a special kind of discriminative stimulus. Other behavior analysts have since noted that a verbal instruction is not a stimulus whose presence signals the current contingencies of reinforcement. Typically an instruction causes certain other stimuli to function discriminatively. "When the tip of the rod jiggles, pick the rod up and set the hook." The effect of this instruction is to cause rod jiggles to function discriminatively for the hook-setting response. "If you see a narrow path with fresh droppings on it, build your blind in a tree close to the path." This instruction causes certain types of paths to function discriminatively for the act of building a hunting blind. Thus, Skinner's analysis has been widely rejected by behavior analysts. It seems that a new functional category different from anything encountered in the animal laboratory is at work here (Schlinger & Blakely, 1987).

From a behavioral standpoint, the most important property of a verbal instruction is its ability to alter the function of other aspects of the organism/environment system. A laboratory technician tells a human subject that points will be redeemable for money at the end of the experimental session. The technician thereby causes any increase in the numeral on the counter to function as a reinforcer. Or perhaps the technician tells the subject that the red light indicates that every 100 button presses will result in a point. She thereby causes the red light to function as a discriminative stimulus for a fixed-ratio 100 schedule of reinforcement. And so on.

One interesting line of research compares the effect of such verbal instructions with the effect of contingencies of reinforcement. It was discovered that instructions facilitate the acquisition of appropriate responding, but at the same time, cause responding to be insensitive to changes in contingencies of reinforcement (Kaufman, Baron & Kopp, 1966). Vaughan (1987) summarizes the subsequent twenty years' worth of findings as follows:

> Experimenter instructions facilitate stimulus control but are likely to establish insensitivity to changes in contingencies unless there are conspicuous consequences (i.e., punishment) for following outdated or inaccurate instructions. Moreover, if subjects are shaped to respond in a certain way rather than instructed, they show greater sensitivity to changes in the experimental contingencies. (p. 110)

In other words, verbal instructions can bring about rapid acquisition of a certain skill (e.g., appropriate responses to contingencies of reinforcement), but at the cost of reducing the subject's readiness to adjust behavior to environmental change (e.g., to unannounced changes in the contingencies of reinforcement). Clearly, this finding has applications to pedagogy, clinical psychology, and a number of other areas. Indeed, it has an almost oriental resonance to it, with obvious similarities to the wisdom of the Zen master who will not tell his student what to do, but instead provides only the experience that determines what act is appropriate. Although it is more difficult to acquire wisdom this way than on the basis of verbal instruction, it appears ultimately to be more valuable to take the more difficult route.

Function Altering Stimuli. An experimental analysis of behavior includes a specification of the physical circumstances that induce certain aspects of the organism/environment system to function in a certain way. The animal is kept at 80% of its free feeding weight to insure that access to food is reinforcing. It receives differential reinforcement for successive approximations to lever presses until it acquires the target response. It is exposed to a correlation between a light's being on and a certain contingency of reinforcement until the light functions discriminatively with respect to the contingency. These procedures or operations do not define the relations they induce. They are just standard recipes for producing functional categories.

These recipes are quite the opposite of broad causal principles. They are often quite specific to a given species: rats learn to discriminate one sort of stimulus, pigeons another, dogs yet a third, etc. These recipes specify the conditions that cause some part of the organism/environment system to function a certain way. They do not use functional concepts to identify these causes. Instead, they use physical concepts (Killeen, 1987). Recently, however, some behavior analysts have begun introducing functionally defined categories of function altering conditions (Schlinger & Blakely, 1987).

Obviously, instructional stimuli can have function altering effects, but the study of function altering stimuli is not limited to verbal stimuli. Michael (1982), for example, has proposed the term *establishing stimulus* for a stimulus that causes another stimulus to become reinforcing. This concept fills a gap in operant theory that Skinner tried to bridge with (what else?) the discriminative stimulus. But as Michael's careful analysis demonstrates, the discriminative stimulus and the establishing stimulus function quite differently. A discriminative stimulus alters the rate of responding because it has been correlated with the enhanced effectiveness of responding in bringing about some reinforcing event. An establishing stimulus, on the other hand, alters the rate of responding because it confers reinforcing properties on some event which responding typically brings about.

Michael gives the following example to separate the two.

> Consider a food-deprived monkey in a chamber with a chain hanging from the ceiling and a retractable lever. Pulling the chain moves the lever into the chamber. Pressing the lever has no effect unless a light on the wall is on, in which case a lever press dispenses a food pellet. . . . We would expect a well-trained monkey ultimately to display the following repertoire: while the wall light is off . . ., the chain pull does not occur . . ., even though it would produce the lever. . . . When the light comes on . . ., the monkey pulls the chain and then presses the lever . . . and eats the food pellet that is delivered. (p. 153)

The basic point is that even when the wall light is off, a pull of the chain will produce the lever, so the wall light is not a discriminative stimulus indicating that chain pulls are effective. Instead, what the wall light does is cause the availability of the lever to become reinforcing (because it signals the effectiveness of lever pressing in producing food). Thus, the wall light has two functions: as a discriminative stimulus of the contingency between lever pressing and food, and as an establishing stimulus with respect to lever-availability. In the latter role, it causes an increase in responses (chain pulls) that produce lever-availability. Note that the establishing stimulus covers more or less the same ground that is covered in philosophy by the concept of an instrumental (as opposed to an intrinsic) good. In Michael's example, lever-availability is an instrumental good, whereas food availability is an intrinsic good. Obviously, the concept of establishing stimulus adds an interesting dimension to the analysis of behavior.

Sidman (1986) has pioneered a related conceptual innovation by extending the traditional three term contingency to four, five, and even six terms. Consider the extension to a four term contingency. Imagine a pigeon in a three key chamber. The key on the left is sometimes (but not always) illuminated with a red light, the key on the right sometimes (but not always) with a green. A standard three term contingency would be: pecking on a key when it is illuminated with green light yields reinforcement. Pecking the key is the operant response, green illumination of the key is the discriminative stimulus, and delivery of seed is the reinforcer. Now suppose we complicate the contingency by making this three term relationship contingent upon some fourth term. For example, pecking an illuminated green key is effective only when the middle key is illuminated with a triangle, and pecking a red key is effective only when the middle key is illuminated with a square. The figure on the middle key thus functions as part of a four term contingency: If the middle key is a triangle, then if a key is illuminated with green light, then if the pigeon pecks on that key, then seed will be delivered; and if the middle key is a square, then if a key is illuminated with red light, then if the pigeon pecks on that key, then seed will be delivered. Under this four term contingency, the figure on the middle key has the function of altering the function of red and green illumination. When it is a square, red comes to function as a discriminative stimulus; when it is a triangle, green comes to function as a discriminative stimulus. This is a four-term contingency. Notice that by the very nature of this contingency, square and triangle have the function of causing red and green to function discriminatively.

Continuing this theme of extending functional descriptions beyond the traditional ones, we turn to Sidman's interesting concept of *equivalence classes*. An equivalence class is a special type of four-term contingency. Returning to our initial example of a three key chamber, where triangle on the middle key signals that green is discriminative for pecking and square signals that red is discriminative for pecking, suppose we trained a pigeon on this contingency. Now suppose we add to our complex contingency a reversal of roles: if the middle key is either red or green, and if the left and right hand keys are illuminated with squares or triangles; then red on middle signals the effectiveness of pecks on square, and green on middle signals the effectiveness of pecks on triangle. Under this contingency, triangle is equivalent to green, and square is equivalent to red. There are thus two equivalence relations in the contingency.

For an animal that has acquired this equivalence relation, the functions of the third and fourth terms are identical: triangle signals the effectiveness of green, and green signals the effectiveness of triangle; square signals the effectiveness of red, and red signals the effectiveness of square. Such equivalence relations, once acquired, can interact with one another to produce novel behavior. For example, if triangle is equivalent to green and green is equivalent to chamber-light-off, then (even if the animal has never experienced green in the presence of chamber-light-off) green will for this animal be equivalent to chamber-light-off. So any response for which green functions as a discriminative stimulus will also be behavior for which chamber-light-off functions as a discriminative stimulus.

As a result, the acquisition of one discriminated operant will bring with it the acquisition of another, even though there has never been an occasion to experience the latter. The capacity to form equivalence classes is sometimes cited as an important source of novel yet appropriate behavior. Interestingly, only human beings have been observed to have this capacity (Sidman, Rauzin, Lazar, Cunningham, Tailby & Carrigan, 1982; Hayes, 1989). This seems to draw a qualitative line between animal and human behavior and to locate with some precision the limits of the strategy of extrapolating from animals to human beings. Thus, this concept not only holds out the promise of expanding the ability of behavior theory to account for novel behavior, it also represents the first clear case of a functional capacity that cannot be studied through animal research.

Summary. Upon occasion, Skinner implied that the three-term contingency provides an adequate basis for explaining the whole of human operant behavior (e.g., Skinner, 1957). Contemporary behavior analysts, however, view aspects of human learning as different in kind from animal learning, and see a need to explore qualitatively different functional relations than one finds in animal research. It is no long accurate to say that behavior analysis is confined to the strategy of studying animals and then extrapolating to human beings.

II

Let us return to our point of departure, which was the anomalous survival of the operant program. If we were to summarize our attempt thus far to explain this survival, we might say that the successes of the operant program are the product of an insight into what properties yield a causal analysis of intentional behavior.[3] If the organism/environment system fits certain functional categories, then certain physical properties of the system (e.g., contingencies of reinforcement) will control others (e.g., rate of responding). So if one defines stimulus and response functionally, and if one asks why (over an extended period of time) a certain quantitative aspect of the response takes a certain value, and if one permits the independent variables to refer to the organism's environment over an extended period of time, then one finds causal regularities that relate environmental causes to behavioral effects (Staddon, 1973; Rachlin, 1974; Hinson, 1987; Malone, 1987; Baum, 1989).

On this view, the behavior analytic program has survived because it accomplishes something that cognitive psychology does not. At the same time, its growth has been limited by the fact that cognitive psychology accomplishes something that operant psychology cannot. This would seem to imply that the simultaneous rise of cognitive psychology and behavior analysis was actually a single historical process having to do with the replacement of the dominant approach of the preceding period. This period is sometimes referred to as the era of learning theory, and one might characterize the major figures in academic psychology of the time as believing that a single methodology could solve two quite different research problems: the problem of defining the principles of learning, and the problem of explaining these principles on the basis of simpler processes. As this point of view began to break down in the early 1950's, two specializations arose almost simultaneously, each making significant

---

[3]Equivalently, it is based upon some assumptions about which behavioral predicates are projectible (Goodman, 1955).

progress on one of these two problems. Behavior analysts made rapid progress on the problem of defining the principles of learning; and cognitive psychologists made rapid progress on the problem of accounting for these complex patterns of learned behavior on the basis of underlying cognitive processes.

The New Molarism. In brief, our explanation for the success of the operant program is that it has found a way to analyze the organism/environment system into entities that enter into causal relationships with one another. Still, one wants to ask, "Is that all there is to the operant program?" As Skinner himself said, even a complete catalog of the effects of each conceivable schedule of reinforcement (something far beyond our reach) would fail to constitute a theory of operant behavior. So even granted that behavior analysis has the ability to discover things that cognitive psychology would necessarily overlook, the question nonetheless arises of whether it has the ability to relate its discoveries to one another in the form of a theory? If not, then one may ask whether it amounts to little but a special methodology, an efficient way perhaps to generate a technology of behavior, but not a scientific discipline--or even subdiscipline. One of the surprises in operant psychology's history--and one reassuring indication that this tradition continues to maintain creative contact with a subject matter--is the unexpected way in which it has provided the basis for an answer to this question.

Our standard examples of causal regularities at the molar level have been the fixed-interval scallop and the fixed-ratio stair step. Recent research, however, has centered on the matching law. The matching law emerged from research on concurrent schedules of reinforcement. Although concurrent schedules are a natural extension of the experimental techniques described in Skinner (1938), he never showed much interest in them. He did, however, introduce the topic in the late 1940's, and included brief reports on them in Skinner (1950) and Ferster and Skinner (1957). He assumed they were derivative of more elementary behavioral processes, and the central goal of his own research was to understand the latter. For purposes of pursuing this goal, even single schedules were proving difficult to handle. So it seemed to him unnecessary (even perhaps irrational) to devote extensive research efforts to the study of a higher level of complexity.

His own theoretical interests continued to focus upon cumulative records of single schedules of reinforcement. Such a cumulative record is a curve on a set of coordinates, and therefore is equivalent to a function that could be expressed mathematically. But what is known about single schedules can for the most part be adequately expressed by describing the shape or angle of inclination of the cumulative records that the various schedules produce. Therefore, research on single schedules has, despite the inherently quantitative nature of its subject matter, seldom required extensive use of mathematics. The topic of concurrent schedules, however, led almost immediately to the introduction of quantitative formulations. Findley (1958) was the first to analyze concurrent schedules in a quantitative manner, but it was Herrnstein's (1961) formulation of the matching law that introduced the basic pattern of the quantitative analysis of behavior.

The question Herrnstein addressed was this: over an extended period of time, what proportion of the subject's behavior on a concurrent schedule will be devoted to each alternative? The meaning of "extended period" is vague, but in practice it is usually several hours of steady state behavior. One intensively studied case presents the subject with two variable-interval schedules on separate keys. Suppose we have a VI-60 second schedule on the left key and a VI-90 second schedule on the right. As the subject works on one key, the other schedule continues to run, meaning that the longer spent on one key the higher the likelihood that switching keys would result in delivery of a reinforcer.[4]

Under these conditions, how will the subject apportion its responses between keys? The answer, according to the matching law, is that (over an extended duration) the proportion of responses at each key will match the proportion of reinforcement received at each key. Algebraically, this is expressed as:

---

[4]The arrangement typically imposes a cost for switching from one response to the other to prevent 'superstitious' alternation between keys every couple of responses. This cost comes in the form of a changeover delay, usually of a few seconds, during which no reinforcer can be delivered.

$$\frac{B\,(1)}{B\,(1) + B\,(2)} = \frac{R\,(1)}{R\,(1) + R\,(2)},$$

where B (1) and R (1) refer to behavior frequency and reinforcement frequency on the first key, and B (2) and R (2) refer to behavior frequency and reinforcement frequency on the second key.[5]

The causal pattern known as the fixed-interval scallop simply tells us the shape of the curve that will appear on a cumulative record when the animal is under the control of a single fixed-interval schedule. The matching law, on the other hand, tells us algebraically how the quantity of one variable is a mathematical function of the quantity of another. This relationship has been confirmed for a large number of reinforcers and a wide range of species, including human beings in laboratory as well as natural settings (see McDowell, 1988, for a summary). Furthermore, it has been extended to additional independent variables, including the amount of reinforcement provided by each successful response (as opposed to frequency of reinforcement), and the immediacy with which reinforcement is delivered following a successful response. These variables can be combined into a single equation that predicts choice behavior when several independent variables are changed at the same time. For example, the expanded equation successfully predicts how much additional food must be provided with each reinforcement in order to offset a decrease in the immediacy of reinforcement (Rachlin & Green, 1972).

Operant psychology has thus found a systematic kind of order at the macro-molar level. Earlier discoveries, such as the empirical principle that fixed-interval schedules generate scalloped cumulative records, were robust, but were not easily related to one another. They were a

---

[5]Davison and McCarthy (1988) refer to this equation as the strict matching law to distinguish it from subsequent modifications. These modifications track the history of modifications of Boyle's Law, including the introduction of constants and new variables.

collection of facts, not a theory. With the matching law, however, operant psychology has entered the realm of theory. For the first time one can compare with some precision the relative contribution of two or three different factors to intentional behavior. Although Skinner did expect eventually to discover this type of knowledge, he expected to find it at the molecular, not the macro-molar, level. The matching law thus provides a novel and unanticipated exemplar of the type of knowledge pursued by operant psychologists.

Research on concurrent schedules appears to be a step in the direction of the world outside the laboratory. An experimental chamber fitted with a single lever on a fixed-interval schedule has few parallels in the natural world, for organisms almost always have more than one productive response available to them.[6] For example, the bird foraging a patch of grass for seeds is capable of flying off to another patch--and sooner or later does. But when? A two lever box with different contingencies for each lever is an abstract model of such an environment, in which several competing sources of reinforcement, each with its own contingencies, are available. Thus, one finds the matching law being applied to natural behavior in a way that single schedule research was not--e.g., to animal foraging (Staddon, 1980; Commons, Kacelnik, & Shettleworth, 1987), and to adult human behavior in complex social settings (Myerson & Hale, 1984; McDowell, 1982, 1988).

This does not, however, mean that the major benefit of the matching law is an increase in prediction and control outside the operant chamber. One of the seldom appreciated aspects of Skinner's theoretical work is his emphasis upon the goal of understanding the behavior of the whole organism. This is something that cognitive psychology has not been notably successful at doing. Instead, it tends

---

[6]Indeed, Herrnstein (1970) has argued that all behavior is choice behavior. Even the animal in the chamber with a single key can cease responding for a while and scratch itself or explore the corner of the chamber. It is as if there were a second key in the chamber with its own schedule of reinforcement, but the experimenter does not have this key hooked up to any recording device. Thus, there is a sense in which behavior on a single schedule can be viewed as the outcome of performance on a concurrent schedule, but we have only one of the two records.

to divide the organism into subsystems which operate in a semi-autonomous manner, and to postpone indefinitely the integration of these subsystems into a whole organism. There is a sense, however, in which the framework provided by behavior analysis affords, even in its early stages, an understanding of the behavior of the whole organism. When a recurrent pattern of modulations in the frequency of some response is attributed to the schedule by which a reinforcer is delivered, this says something about the net effect of the functioning of the entire organism. There is a level of understanding here which is not necessarily equivalent to an understanding of the operation of the various subsystems which make up the organism.

A similar kind of understanding is provided by the matching law and its many variants, which say that over an extended period, an organism maintains a certain objectively definable relationship with the environment. Whether this relationship can be defined as optimizing the value of certain parameters, or whether some other mathematical function is involved, is a hotly debated issue. Whatever the outcome, though, the mathematical theories emerging from the matching law tradition provide a profound (although perhaps mistaken) interpretation of what behaving organisms, including human beings, are doing. Even when these theories do not contribute to the prediction and control of behavior outside the experimental chamber, they provide a form of understanding that has the potential to unify our conception of the learned behavior of organisms.

Relationship to Cognitive Theory. The demise of radical behaviorism creates a new context within which to conceptualize the relationship between behavior analysis and cognitive psychology. To this end, it is useful to follow Cummins (1983a) in distinguishing between two types of scientific theory. On the one hand, there are transition theories, which "explain changes of state in a system as effects of previous causes" (p. 1); and on the other, there are property theories, which "explain the properties of a system not in the sense in which this means 'Why did S acquire P?' or 'What caused S to acquire P?' but, rather, 'What is it for S to instantiate P?', or 'In virtue of what does S have P?'" (pp. 14-15). Transition theories explain by subsumption under causal principles. Property theories explain by analysis--by showing that something having certain components organized in a certain way is "bound to have the target property" (p. 17). Discovering explanations of the transitional type will add to our understanding of what causes a given event to occur; discovering explanations of the property type will add to our understanding of the processes that underlie an event. The first type of explanation is non-reductive--it simply specifies as accurately and as comprehensively as possible what causes what--whereas the second type of explanation shows how the causes generate their effects.

A non-psychological example of these contrasting explanatory types would be the molar gas laws and the kinetic theory of gases. The kinetic theory accounts for the properties and dispositions formulated in the molar laws. The molar laws attribute certain dispositions to gases, among which is the tendency for an increase in temperature to cause an increase in pressure. The kinetic theory, on the other hand, analyzes the property of temperature as mean kinetic energy of gas molecules, and explains the molar relationship between temperature, volume, and pressure by applying Newtonian physics to an idealized model of what gases are. Transition theories and property theories answer different questions. Consider the ability to distinguish between the grammatical and the ungrammatical sentences of English. This is an ability possessed by competent speakers of English. A transition theory would attempt to specify the factors that cause a person to acquire that ability. The theory would thus identify the conditions that increase or decrease the probability of becoming competent in English. A property theory, on the other hand, would target a different question. It would attempt to explain how the ability is embodied. A step in the direction of such an explanation might be to analyze the complex ability into a set of simpler abilities which can be programmed in a certain way so that anything having the simpler abilities related to one another in accordance with the program would thereby display the complex ability.[7] Such a decomposition provides

---

[7]Cummins calls such accounts "functional analyses." Skinner has used the term *functional analysis* in a quite different (indeed, opposite) sense. Fortunately, what Skinner means by *functional analysis* can be conveyed by the term *causal principle*. I have taken advantage of this fact, and used *causal analysis* or *causal principle*

a partial explanation of how a property is instantiated.[8]

The molar gas laws formulated a quantitative relationship among volume, pressure, and temperature; then the kinetic theory of gases explained this relationship on the basis of a hypothetical underlying process consisting of the random motions of the molecules that make up the gas. The emergence of the matching law as an important topic of research suggests that operant psychology may stand to cognitive psychology as classical thermodynamics stands to the kinetic theory of gases, and that one of the tasks of cognitive psychology will be to explain the matching law.[9]

Not all property analyses require ontological descent. Consider the chemical analysis of photosynthesis. The molar regularity is that carbon dioxide, water and sunlight go in; oxygen and carbohydrates come out. The problem is to analyze the intervening process, revealing the contribution of chlorophyll and various enzymes to the outcome. To do so there is no need to descend to the level at which a chemical compound is viewed as a stable configuration of atomic nuclei bound together by shared electrons. It is enough to analyze photosynthesis into a sequence of simpler chemical reactions. In practice, of course, the two levels are often mixed, thereby providing a fuller understanding of the process. But the chemical analysis of photosynthesis can proceed without ontological descent to the level of atomic nuclei and electrons.

Skinner's theory of schedule effects was based on the thesis that a

---

wherever Skinner has used the term *functional analysis*, thereby reserving the term *functional analysis* for the concept Cummins would use the term to convey. Notice that Cummins contrasts functional analysis (of underlying processes) with (merely) causal accounts (which are what Skinner would refer to with the term *functional analysis*). Probably the only reason this attempt by two individuals to use the same term to mean exactly opposite things has not caused significant confusion is that few people who follow the one's usage are aware of the other's.

[8]See Bechtel & Richardson (1993), however, for a discussion of some limitations and pitfalls of the decompositional approach to property reduction.

[9]The analogy between the matching law and classical thermodynamics has been discussed by Marr (1984, 1989).

similar strategy would suffice for the analysis of behavior. Molar behavioral regularities could be analyzed into a complex interaction of simpler behavioral regularities. Although certain inferred entities might be needed (e.g., interoceptive stimuli), these entities would still be behavioral (as opposed to mental or physiological). Eventually ontological descent would be necessary, but Skinner thought it would proceed directly from elementary behavioral entities and processes to physiological entities and processes. Mental concepts would not enter into the analysis.

As an analytic strategy this was impressively parsimonious, but ultimately unworkable. The past twenty-five years of operant research is difficult to reconcile with behaviorism's campaign against mental entities. This does not, however, mean that behavior analysis must abandon the goal of discovering environment-to-behavior regularities. One may acknowledge the appropriateness of mentalism for one scientific purpose while pursuing a quite different purpose for which mentalism is not necessarily the best approach. If one seeks a property theory of behavior, then mentalistic concepts may indeed be indispensable; but if one seeks a subsumptive account of behavior, then perhaps they are not.

Despite sweeping dismissals by Skinner, some behavior analysts have shown an interest in cognitive processes. A few, for example, have speculated about the cognitive processes that underlie schedule effects (e.g., Shimp, 1984; Staddon, 1983). Furthermore, there has been an explicit and formal acknowledgment in the *Journal for the Experimental Analysis of Behavior* that it is legitimate (given sufficient empirical support) to attempt to explain behavioral principles on the basis of underlying cognitive processes. In a pair of brief but significant editorials, Nevin (1980) first states and Hineline (1984) later reaffirms that the journal now accepts articles that offer cognitive explanations of behavioral phenomena. What qualifies an article for publication is the rigor of the data and the ability of the proposed explanation to account for them, not the nature of the entities referred to in the explanation. Hence, there is now within behavior analytic circles an acknowledgment of the behavioral evidence for the existence of cognitive processes.

A Division of Labor. For any given learned pattern of behavior there are two quite different types of explanation: (1) a causal account of why the pattern arose, and (2) a reductive account that shows the pattern to be the result of a more basic process. Our account suggests a division of labor between operant and cognitive psychology that parallels this distinction. Operant psychology provides the first type of explanation, cognitive psychology the second. This suggestion may not win universal agreement, but at least (at a given level of behavioral description) it makes a clear assertion. There are, however, many levels of description. For example, there is not only the disposition to respond in a certain way in the presence of a certain stimulus, but a disposition to acquire this disposition. And this second level of disposition creates some possibilities for confusion.

Take the ability to distinguish between the grammatical and ungrammatical sentences of English. Chomsky (1980) has proposed that a system of rules and representations underlies this ability. Skinner (1957) has proposed an alternative analysis. Both analyses attempt to give accounts of the second type. There is however another question one can ask about the ability to distinguish between grammatical and ungrammatical sentences of English. One can ask how this ability arises. Here again, there are two different types of account one might seek. On the one hand, there is an account that identifies what causes a child to acquire this ability. The answer might include factors such as the practical benefits that accrue to language use in the child's social environment, the relative simplicity of sentences used in the presence of the child, etc. On the other hand, one might ask what process underlies the child's ability to acquire this ability. The answer to this second question is completely different in kind.

Chomsky has proposed that the ability to acquire language is itself underlain by a system of rules and representations. So he posits a system of rules and representations to underlie the process of acquiring a system of rules and representations. Such a theory does not compete with the theory that the event of acquiring a language is caused by the practical benefits of language use. Hence, evidence that language is not acquired unless there is some practical payoff is not evidence that we do not need to posit an underlying system of rules and representations. The two are complementary explanations that answer different questions.

A Different Kind of Hypothesis. Operant psychology is sometimes thought to be totally committed to a Baconian method that eschews the use of hypotheses. We have seen however that Skinner himself worked on a grand hypothesis (his theory of schedule effects) that could not have been derived in Baconian fashion from his experiments. Skinner did, however, attempt to do without hypothesis in his search for causal principles of the sort that his hypothesis was supposed to explain. And since operant psychology has more or less given up the quest for a behavioral explanation of such casual principles, one might think it has no further need for behavioral hypotheses. This, however, is not the case. The matching law moves the analysis of behavioral regularities beyond the mere cataloging of facts, and into the realm of hypothesis. In a sense, behavior analysis has recapitulated a point about scientific method that was made in the middle of the nineteenth century by John Stuart Mill (1806-1873). Drawing upon Francis Bacon's account of scientific logic, Mill (1865) defined four methods by which causal regularities are discovered: the method of agreement, the method of difference, the method of concomitant variations, and the method of residues. None of these methods requires the use of hypothesis, and Mill is often portrayed as claiming that all causal regularities can be discovered by one or the other of these methods. In fact, however, he claimed only that certain types of causal relationships could be discovered in this manner. Others, by their very nature require use of the hypothetico-deductive method.

Of special interest to us is the fact that he did not draw the limits of the logic of discovery by reference to hypothetical entities. Obviously, any principle making reference to such entities will require the use of imaginative hypotheses. But Mill's discussion shows that not all unavoidable uses of hypotheses make reference to such entities. In particular, Mill argued that certain types of multiple causation obey laws that cannot be discovered through the use of his four methods. An example is the case of motion caused by multiple impressed forces. The net effect of two or more impressed forces can be derived from composition of the effects of the individual forces. Such an effect, however, could be the outcome of infinitely many different combinations of impressed forces.

> Mill concluded that his inductive methods were unavailing in cases of the Composition of Causes--one cannot proceed inductively from knowledge that a resultant effect has occurred to knowledge of its component causes. For this reason, he recommended that a "Deductive Method" be employed in the investigation of Composite Causation. (Losee, 1980, p. 153)[10]

In brief, Newtonian physics could not have been discovered through the Baconian methods.

This provides an interesting comparison with behavior analysis. The Newtonian treatment of multiple causation resembles the matching law's treatment of the various factors that control choice behavior. Just as there are infinitely many different combinations of impressed forces that could account for a given motion, so also are there infinitely many different combinations of rate of reinforcement, delay of reinforcement, and amount of reinforcement that could account for a given allocation of responses. Thus, the matching law describes the type of causal regularity that Mill saw was beyond the reach of Baconian methods. And true to form, research in the matching law tradition proceeds by hypothesis (Davison & McCarthy, 1988). Yet, like Newton's laws of motion, the matching law makes no reference to

---

[10]One of the stumbling blocks to interdisciplinary works such as the present one is the fact that technical terms sometimes take on distinct, but related, meanings within distinct, but related, disciplines. The concept of *induction* is a case in point. Contemporary logicians use this term to denote the genus of non-deductive (ampliative) inferences; but historians of science (and a number of other academic groups) use the term to denote a specific type of non-deductive inference namely, the type of inference that proceeds from the specific to the general. Mill's methods of scientific discovery were an early attempt to formalize the principles governing this species of inferences.

All science is based upon inductive inference in the generic sense. Only certain limited portions of science, however, can proceed exclusively on the basis of inductive inferences in the specific sense. Skinner (1950) seems to imply that behavior analysis is an instance. Ferster and Skinner (1957) provide a counterexample.

hypothetical entities. Thus, contrary to what one might have expected, it was Skinner who introduced hypothetical entities into behavior analysis, and his students who discovered how to develop genuine theories without them.

CHAPTER TWELVE

OPERANT PSYCHOLOGY WITHOUT BEHAVIORISM

Many behavioral psychologists now concede that reference to cognitive mechanisms is necessary to provide explanations of behavioral regularities, but few would concede that reference to such entities is necessary to formulate the regularities themselves. They believe that if one studies large units of behavior rather than discrete building blocks, if one focuses upon complex relational stimuli rather than upon simple non-relational stimuli, and if one attends to the equilibrium that emerges after prolonged exposure to a stimulus rather than to moment-by-moment changes, then one can formulate simple behavioral principles that connect complex environmental causes to complex behavioral effects.

This has sometimes been taken to mean that "virtually all significant voluntary human actions can be understood in terms of their past relations to rewards and punishments" (Schwartz & Lacey, 1982, p. 15), or that human action is under the control of contingencies of reinforcement (Skinner, 1971). If we ask whether contemporary behavior analysts assume voluntary (operant) behavior to be under the control of environmental factors such as these, the answer is yes and no. Yes, they think operant behavior is under the control of environmental factors, but no, they do not think we can limit the list of such factors to the familiar three-term contingency of discriminative stimulus, operant response, and reinforcer. For one thing, they know that the effect of contingencies--even in a highly constrained environment such as the conditioning chamber--can be significantly altered by the availability of reinforcement elsewhere. One gets quite different cumulative records for identical schedules of reinforcement under open and closed economies. Furthermore, there are environmental relationships besides the three-term contingency that play an important role in determining operant behavior, including instructional stimuli, establishing stimuli, equivalence classes, and four-, five-, and six-term contingencies.

So even though behavior analysts assume operant behavior to be under environmental control, they do not take this control to be exercised solely by (three-term) contingencies of reinforcement, but instead take the search for the sources of control to be their central scientific problem. New solutions to this problem are constantly being proposed--but always within the context of the working hypothesis that operant behavior is somehow under environmental control. The latter hypothesis, and not the assertion that behavior is under the control of contingencies of reinforcement, is the central tenet of behavior analysis. Therefore, although behavior analysis is committed to an environmentalism of sorts, this environmentalism does not contrast with nativism. Instead of assuming that all behavior is learned, it assumes that all learned behavior obeys environment-to-behavior causal principles (T. L. Smith, 1983).

There is no quarrel with ethologists who emphasize the differences to be found in the unlearned behavior of different species. Furthermore, inasmuch as some of the more recently discovered causal relations between learned behavior and the environment have not been observed in any animal other than human beings, it can no longer be said that behavior analysis is committed to the view that the principles of human learning can be extrapolated from the behavior of rats and pigeons. Perhaps if the three-term contingency had turned out to be an adequate basis for explaining the whole of human operant behavior, then the main function of the experimental analysis of human behavior would have been simply to set the value of a few coefficients that vary from species to species. At one time, Skinner seems to have held such a view (Skinner, 1957), but later he appears to have abandoned it (Skinner, 1966). In any event, most behavior analysts are inclined to view parts of human learning as differing in kind from animal learning--a development that has helped to clarify the aspirations of the program.

I

The relationship of operant psychology to cognitive psychology has become a complex blend of complementarity and competition. The two complement one another in this respect: operant psychology formulates causal principles about the environmental causes of learned behavior; cognitive psychology provides an account of the mechanisms that underlie these environment-to-behavior causal relationships. On the other hand, the two approaches are in competition with one another in this respect: cognitive psychology (a version of which is the economist's theory of rational choice) attempts to provide subsumptive explanations of ordinary human behavior, and so does operant psychology. Operant psychology however has two comparative advantages in this competition: one is epistemological, the other logical. Due especially to the latter, it has a virtually insurmountable edge in this domain.

Operant Psychology's Comparative Advantages. The first of operant psychology's comparative advantages derives from the fact that behavioral units have quantitative properties that can be measured directly (i.e., without making substantive assumptions about the quantitative values assigned to other psychological states). Given that a certain stimulus is a reinforcer, for example, it is a rather straightforward matter to measure the frequency with which the reinforcer is delivered. And frequency of delivery is one of the variables that enter into quantitative principles. The quantitative aspects of cognitive entities, on the other hand, can be measured only indirectly by making certain assumptions about the quantitative values of other psychological states. The difficulties posed by such circularity for the pursuit of causal regularities are perhaps not insurmountable, but they are real. Hence, there is an epistemological advantage to the behavioral approach that makes it easier to confirm and refine precise quantitative principles.

A second, and more important, advantage conferred by behavioral concepts is that they are capable of defining regularities that satisfy the principle of compositionality. Hence, when more than one behavioral cause exerts influence over the same variable, it is (at least theoretically) possible to calculate the net effect. The matching law is the primary exemplar. It tells us, for example, the degree to which a delay of reinforcement for a given response will tend to shift responding to alternative responses. It also tells us the degree to which an increase in relative frequency of reinforcement for a response will have the opposite effect. The causal tendencies imparted by changes in delay and frequency of reinforcement maintain their validity when they interact, and so they can be composed into a net result that tells us how much of an increase in frequency of reinforcement will be necessary to offset a given increase in delay of reinforcement. .

It is not likely, however, that cognitive principles can be composed in this fashion. The reason is as follows:

> Suppose that human action is due to underlying states of belief and desire. The effect of any such state is always a function of an indefinite number of other states of the system. So the behavioral consequences of a given belief or desire can be neutralized or redirected by beliefs or desires elsewhere in the system. Hence, there are no cognitive regularities saying such-and-such cognitive state tends to increase the likelihood of such-and-such action. Q.E.D.

Unless one has a total description of the subject's beliefs and desires, prediction is unreliable and explanation is incomplete. This is not just a matter of approximating the truth, because an ignored state could completely reverse the tendencies of other factors. Hence, there are no valid cognitive regularities that tell us the behavior to be expected under a given condition, so long as that condition includes less than a full description of the subject's beliefs and desires.

What, for example, is the effect on behavior of a belief in immortality? It depends on what else a person believes and desires. Does he believe that everyone is immortal, or does he believe that some are but others are not? Does he believe that there are different kinds of immortal life, and if so, does he believe that some are better than others? Does the better sort of immortal life come only to those who act a certain way, or does it come to someone regardless of how he acts? Or is immortality something inherently undesirable, something to be avoided, something to be delivered from? The answers to these and indefinitely many other questions have a bearing on what effect a belief in the afterlife will have on behavior. Folk psychology, of course, offers a passable account of ordinary behavior by assuming (implicitly) that the beliefs and desires of a subject fall within a certain range. Typically, this range is specified by factors such as the age, gender, role, and culture of the subject, and is supplemented by the assumption that the subject is normal. The main function of such categories is to delimit the beliefs and desires a subject may be expected to possess. Within such a context, common sense can predict and explain certain forms of behavior. Thus, a Christian can recognize the behavioral symptoms of a fellow Christian's faltering belief in immortality, but might be baffled by the effects of disbelief on the behavior of a Buddhist.

Careful students of human cultures are often skeptical about the possibility of accounting for voluntary behavior by subsuming it under principles that apply to any culture at any time. Perhaps this is because they assume (correctly, it seems) that there are indefinitely many beliefs and desires that human beings are capable of acquiring, and they see the consequences of this conclusion for folk psychology. Conversely, those who are unfamiliar with the diversity of cultural systems are often those who have the greatest faith in the possibility of cross-cultural principles of human behavior.

Actually, the situation is even more complex than this, for reasons that have been set forth by Jerry Fodor (1983). There seem to be various modular components of the cognitive system that are relatively autonomous (in the sense that beliefs and desires in other parts of the system do not have an impact there). And these modular components seem to be susceptible to valid cross-cultural analyses (e.g., of linguistic universals). But when the topic shifts from the modules to the behavior of the organism as a whole, Fodor suddenly becomes skeptical about our prospects for arriving at principles that have cross-cultural validity. He does not say why, but we may speculate it is for reasons similar to those given above. Unless the total system is highly constrained, it is difficult to see how valid principles of the behavior of the organism as a whole can be formulated in cognitive terms (see also Chomsky, 1975; Davidson, 1980, 1984).

Cognitive psychology appears to be unable to put together (compose) the behavioral tendencies of the organism as a whole. The disposition imparted by a given belief or desire is relative to the subject's other beliefs and desires. Without information about the content of these other beliefs and desires, one cannot say whether a change in a given cognitive state will tend to increase or decrease the probability of a certain action (let alone how *much* it would increase or decrease it). This kind of context sensitivity interferes with compositionality of effects. For example, suppose that in context C the addition of cognitive state A would cause an increase in behavioral effect E, but that in context C the addition of cognitive state B would cause a reduction in effect E. We cannot infer that A and B together in context C would cancel one another out, however, for each cognitive state will itself now become part of the context for the other. State A in context C + B may no longer increase E. State B in context C + A may no longer decrease E. The explanatory principles of cognitive psychology in this sense do not possess compositionality, whereas the principles of behavior analysis do.

Where These Advantages Make a Difference. There are at least two areas where these conceptual advantages seem to make a difference. The first has to do with the analysis of social systems and the second with the analysis of multiple causes.

Operant psychology offers a theory of forces. But a theory of forces, even a quantitative theory of forces, gives only a subsumptive account of events. Property theories, on the other hand, provide a different and especially satisfying form of explanation. Indeed, some philosophers do not consider a subsumptive account to provide an explanation at all; they hold that only a property analysis is truly explanatory (Cummins, 1983b). Operant psychology evidently must renounce the goal of giving a property analysis of behavioral capacities. Does this mean behavior analysis can never provide the kind of insight that comes with property theories? Not necessarily.

If behavioral principles are to provide a property analysis of something, it is more likely to be of social regularities rather than behavioral ones. Indeed behavior analysis seems to offer just the right type of principle to explain certain social patterns on the basis of the interaction of individual patterns of responding. Staddon (1984) notes that the actors in a social system can usefully be conceptualized in classic behavior analytic fashion by viewing the individual actor as defining a pair of functions: a control function mapping stimuli onto responses, and a feedback function mapping responses onto stimuli. When two individuals start interacting in such a way that the control function of one generates the feedback function for the other, this constitutes a social system that can be analyzed behaviorally. The behavior of one individual is the environmental stimulus of the other, thereby setting up dynamic behavioral interactions.

Such a system can amplify small tendencies into large ones. A simple example has been provided by Tinbergen and Tinbergen (1973), who speculate that a certain type of adult/child interaction is responsible for the allegedly high incidence of autism in modern societies. Their hypothesis is that if a child with a timid disposition gets paired with an intrusive adult, a harmful dynamic can ensue. Suppose the intrusive adult abruptly tries to initiate a social interaction such as physical touching or mutual eye gazing. The child, being shy, turns away. At this point, if the adult would remain physically present but passive, the child would eventually make social contact. In traditional societies, according to the Tinbergens, this is typically what occurs. In modern societies, however, adults often react to a child's withdrawal by becoming even more intrusive, which in turn causes the child to become more withdrawn, which causes the adult to become more intrusive, etc. The outcome, say the Tinbergens, is sometimes a child so withdrawn that he or she no longer acknowledges social stimuli--i.e., an autistic child.

My point is not that this theory is correct. (Indeed I am told it is not.) Nothing in the analysis, however, turns on the actual truth of the theory. My point is just that it provides a simple model of the type of property theory one presumably would seek in order to explain (in the strong sense noted above) social patterns or regularities.

This illustration is valuable because it can be described without the use of a quantitative analysis. Staddon (1984), however, notes that it is usually impossible by verbal reasoning alone to predict the effects of deterministic systems when they interact dynamically. He illustrates his point by examining the interaction between an aggressive child and a punishing parent. By formalizing a few assumptions about the causes and effects of punishment (e.g., that the parent increases the rate of punishment when the child increases the rate of rule breaking, and that an increase in the rate of punishment decreases the rate of rule breaking), Staddon shows that the parent/child system falls into a cyclical equilibrium. This type of theoretical activity is well suited to behavior analysis (see Battalio, 1973; Rachlin, 1980; Rachlin, Green, Kagel & Battalio, 1976; Staddon, 1983).

The second problem area where behavior analysis seems capable of making a distinctive contribution is the analysis of multiple causes. A theory of forces, especially a quantitative theory of forces, is Everyman's ideal of a scientific theory. It allows us, under controlled conditions at least, to say with precision what the outcome of conflicting causes will be. Instead of being reduced to the journalistic formula of saying "On the one hand we have this factor, but on the other hand we have that factor, and only time will tell the outcome," a quantitative analysis holds out the hope of meaningful predictions. Of course, even well developed sciences have trouble predicting many natural phenomena. Skinner (1953a) noted that a physicist would be hard pressed to predict the changing temperature of his morning cup of coffee, because there are too many unmeasured factors interacting. And the same would be true of much ordinary behavior, even if we had a well developed behavior analysis of it. Even so, there is a certain kind of understanding that such a theory of forces provides, in that we gain a sense of proportion about the relative contribution of various causes.

Consider the question of the relative control exercised by concurrent contingencies of reinforcement. Does the allocation of behavior under such conditions of choice maximize some quantitative aspect of behavior (e.g., total amount of reinforcement), or does behavior reach equilibrium at some other quantitative point? The quantitative version of folk psychology--i.e., the economic theory of rational choice--has traditionally assumed that human behavior reaches equilibrium at the point that maximizes marginal benefits. One of the surprising features of the matching law (at least on Herrnstein's meliorizing interpretation of it) is that it implies that this is not so. As a result, the matching law predicts that even under ideal conditions (i.e., even after the subject has had enough time and experience for her behavior to reach a stable equilibrium) the allocation of responses will often be irrational in the sense that it does not maximize anything important to the subject (Ainsley, 1985; Herrnstein, 1990).

Consider as a second example the question of the effect of language upon behavior. We have seen that this can interfere with the control that would otherwise be exercised by contingencies of reinforcement. This is a somewhat surprising area for behavior analysts to be working on, since their methods have not been particularly effective in dealing with language use or language acquisition. But we have seen that contemporary behavior analysts are not asking the same kind of question that earlier theorists asked. By comparing the effect of verbal instructions with the effect of contingencies of reinforcement, they have been able to explore the topic of indirect (language mediated) knowledge versus direct (experiential) knowledge.

These last two examples--the use of the concept of concurrent schedules of reinforcement to explore the traditional topic of rationality versus irrationality, and the use of the concept of instructional stimuli to explore the traditional topic of indirect versus direct knowledge--are creative extensions of the behavior analytic tradition into new domains. In both cases, there is an emphasis upon understanding the phenomena in question by arriving at a reasonable judgment about the net effect that various factors have upon behavior, and not simply upon prediction and control. Although some observers see prediction and control as motivating the entire behavior analytic program of research (e.g., L. D. Smith, 1992), this would seem to be more true of Skinner's aspirations for the program than it is for the program itself.

II

Radical behaviorism tried to pave the way not only for scientific progress, but for social progress as well. Skinner held out the hope that behavior analysis could actually save the world by solving problems such as pollution, nuclear war, and overpopulation. Towards the end of his life he may have become pessimistic about what could be accomplished, but he never blamed behavior analysis for his pessimism. The problem, he said, was not lack of scientific knowledge, but reluctance to use it.

He thought some of this reluctance was due to the contingencies of reinforcement that control the behavior of the people who can solve our problems. Corporations will continue to pollute as long as it is more profitable to do so than not to do so. Governments will confront other nations with their nuclear arsenals as long as governments that do so have a better chance of staying in power than those that do not. People in poor nations will continue to have more children than the environment can support so long as having them is the best way to guarantee an income during one's waning years. Behavior analysts can explain these patterns of behavior, but can do little to change them (Skinner, 1983b). There are also, however, philosophical reasons why we fail to take the necessary steps. And these, Skinner thought, need to be acknowledged and refuted. Such was the intent of Skinner (1971).

Skinner's Radical Critique of Freedom and Responsibility. Skinner believed that the ethical and political practices of Western civilization are based upon an unscientific analysis of behavior, and that this analysis has become an obstacle to progress. In particular, our commitment to individual autonomy, responsibility, and freedom is based upon an outdated metaphysics that assumes the existence of an autonomous inner self--i.e., of an uncaused cause. Skinner's argument against the principle of individual responsibility has been summarized by Schwartz and Lacey (1982, p. 13) as follows:

> If it is true that human behavior is the reliable product of environmental events, then responsibility for behavior, whether noble or ignoble, rests not in the actor but in the environmental variables that give rise to the action. If behavior theory succeeds, our customary inclination to hold people responsible for their actions . . . will be replaced by an entirely different orientation. This new orientation is one in which responsibility for action is sought in environmental events. Such an orientation provides a view of the world that will leave no aspect of daily life untouched.

For support, they cite *Beyond Freedom and Dignity* (Skinner, 1971), which repeatedly asserts that if behavior is determined by the environment, and not by autonomous man, then it makes no sense to hold individuals responsible for their behavior.

Although behavior analysts frequently take critics to task for misinterpreting Skinner, it is impossible to fault the critics in this particular instance. If anything is clear, it is that Skinner says behavioral psychology contradicts the principle of individual responsibility. For he thinks the premise that the environment is causally responsible for human action entails that it makes no sense to hold individual persons responsible for their actions. And since he is committed to the premise, he accepts the conclusion.

Similar considerations apply to the principle of freedom. According to him, allegiance to this principle arose within an historical context in which the major obstacle to human progress was tyranny (i.e., illegitimate forms of authority). By encouraging the belief that individual human beings are (and should be) free from external control, the concept of freedom inspired resistance to tyranny. Unfortunately, many people came to believe in human freedom as a metaphysical fact. This was of little importance so long as tyranny remained the major problem. Now however we face problems of a different kind. Advertisers stimulate wasteful consumption, free markets despoil the environment, alcohol and drugs induce dependence. None of these problems is due to the tyrannical power of an illegitimate authority. Quite the contrary, each is due to freedom from legitimate authority. We shall solve these problems only if we are willing to permit some institution (whether government, a *Walden Two* style commune, or something else) to interfere with our freedom. But proponents of freedom object to any interference with human autonomy, and thereby become obstacles to progress (Skinner, 1971).

A Philosophical Puzzle. Philosophers tend to be puzzled by Skinner's argument. The issues of freedom and personal responsibility have been debated by philosophers for centuries, ever since it began to dawn on them that human beings, as part of the natural world, must be part of the same system of causes and effects that we observe in the world around us. But if this is so, and if the causes of our actions can be traced outside us, then how are we to justify the practices associated with the ethical ideals of freedom and responsibility? Does not naturalism undermine the very foundations of these practices? This is traditionally known as the problem of freedom of will versus determinism.

Anyone familiar with the philosophical debate surrounding this problem will be struck by two aspects of Skinner's argument. First of all, Skinner's argument implies that his discovery that behavior has environmental causes somehow resolves this issue. This is odd, because naturalism (the view that human behavior is part of the same system of causes as everything else in nature) has hardly been begging for support. Perhaps if the argument about freedom and responsibility was a matter of assessing the strength of the evidence for naturalism, and the scale weighing freedom and responsibility on the one hand and naturalism on the other was evenly balanced, then the additional evidence from behavior analysis might be crucial to the outcome of the debate. But the debate has not proceeded in this manner.

The philosophical problem is not to make a choice between naturalism and anti-naturalism. Rather, it is to solve a series of puzzles about the implications of naturalism for ethics. So it is odd to find Skinner asserting that since he has discovered that behavior has natural causes, it follows that we should abandon the ideals of freedom and responsibility. Skinner seems to ride rough shod over the questions of philosophical interest.

The second thing a philosopher notices about Skinner's discussion is that he never gives serious consideration to the possibility that ethical principles such as freedom and responsibility are compatible with naturalism. There is quite a substantial body of philosophical literature making a case for this position, and Skinner simply ignores it. Instead, he proceeds as if it is obvious that the discoveries he has made about behavior contradict (and therefore refute) our basic ethical principles. How are we to make sense of Skinner's puzzling behavior?

Radical Behaviorism is the Key. The simplest explanation is that it is not behavior analysis that contradicts the principles of individual responsibility and freedom, but Skinner's philosophy of radical behaviorism. Skinner does not derive his anti-libertarian conclusions from the thesis that behavior has natural causes, but from his property theory of what those causes are. There is no obvious contradiction between traditional ethical principles and the molar principles of behavior analysis. Radical behaviorism, however, requires that molar principles be reduced to more and more basic behavioral capacities until eventually we reach capacities so simple that physiology can take over. This would leave no room for mental causes, and it is this aspect of the analysis--not the environmental determinism--that conflicts with freedom and responsibility.

How so? Briefly, because the point of these social practices would seem to be to induce individuals to come under the influence of certain rational capacities. Roughly speaking, we hold people responsible for their actions and permit them to be free from contrived methods of control because we think doing so will induce them to come under the influence of reason. Granting them freedom will lead them to exercise their rational capacity for choosing how best to use their assets and abilities. Holding them responsible will lead them to take the broad interests of the community into account when weighing their reasons for acting one way or another. Ultimately, the actions resulting from exercise of these rational capacities are (presumably) caused by environmental events. But the justification for our ethical practices rests upon an assumption that these environmental events have their effect on behavior only as a result of mediating cognitive processes. Such, at least, is the standard line taken by compatibilism (e.g., Smart, 1961).

If we did not think such cognitive capacities exist, if instead we thought that behavior is shaped by a process of learning that is analogous to natural selection, then the practices associated with freedom and responsibility would not make much sense. Dennett (1984) argues, quite persuasively it seems to me, that it is precisely the mindlessness of Skinner's account of behavior that conflicts with these practices. But this mindlessness is not a consequence of the discoveries of behavior analysis. It is a consequence of radical behaviorism. And so if we reject radical behaviorism, we can avail ourselves of the usual philosophical arguments on behalf of compatibilism. This is not to say that behavior analysis is committed to, or demonstrates, compatibilism. I do not see how it takes one side or the other on this question. But it leaves open this possibility. And this radical behaviorism did not do.

III

Even if Skinner's radical critique of freedom and responsibility cannot be sustained, behavior analysis nevertheless offers a vision of behavior that has certain broad (though perhaps not revolutionary) social implications. These implications are due to its environmental determinism, according to which the environment is ultimately responsible for adaptive (learned) behavior.[1]

This is the point of view that has informed classical versions of behavior modification and behavior therapy. If someone has acquired a behavioral problem as a result of being exposed to a disfunctional social environment, forget about the mental states and processes that may or may not underlie it. The problem presumably arose as a result of factors that prevailed in the past environment, and it will be solved by changing those factors (cf., Michael, 1985). If a student is showing lack of effort, this could be due to lack of reinforcement (extinction), or to the fact that reinforcement is not closely enough linked to amount of effort, or perhaps there is too long a delay between effort and reinforcement, etc. If juveniles are engaging in crime rather than school work, this may be due to the higher rate of reinforcement for crime over scholarship, or to the immediacy of reinforcement, etc. This intriguing point of view underlies much that is distinctive about the behavior analytic strategy for solving behavioral problems.

Environmental Determinism. Operant psychology assumes that in most cases behavior reaches an equilibrium point that increases (although not necessarily maximizes) the net amount of reinforcement. This in itself is a rather comprehensive vision of human behavior

---

[1]Skinner (1953a) explicitly defends environmental determinism while bracketing the whole issue of whether mental states mediate the relation between environment and behavior. Thus radical behaviorism may imply environmental determinism, but environmental determinism does not imply radical behaviorism.

As some behavior analysts have noted, this vision has obvious affinities with the position of cultural materialism, as defended by Marvin Harris (Lloyd, 1985). The central thesis of cultural materialism is that cultural practices emerge, are sustained, and disappear as a result of the rate of material return on those practices as compared to the rate of material return on available alternatives. One example is the taboo in parts of India on killing of cows. Harris defines this taboo functionally. It is not a given set of beliefs or desires defined in terms of their content, but a system of artifacts and responses that tend to reduce the inclination to kill cows. What he is interested in is the tendency for people to say and do things that reduce the rate at which members of the culture take the lives of cows. His thesis is that this tendency is under the long-term control of the material benefits that accrue to keeping one's cows alive.

He notes that the function of the practice is precisely to keep people from yielding to the temptation to kill healthy cows. If a cow dies from natural causes, there is no taboo against consuming the meat. Furthermore, there is no taboo against taking the life of a bull. Why? According to Harris, it is because cows need to be preserved if a farmer in this particular ecosystem is to succeed. This is especially true during times of famine, when there is a strong temptation to kill one's cow and eat it. The taboo's role is to prevent this from happening. In order to recover once the famine is over, it is necessary for the farmer to have a healthy cow. The cow will not only be a source of food (in the form of milk), fuel (in the form of dried cow dung), and labor (when yoked to the plow), but also a source of its own replacement (in the form of a calf). It is by reference to strategic material contingencies such as these that Harris would explain the taboo. He supports his analysis by examining regional variations on the taboo and relating them to ecological differences in the economics of farming. He argues that variations in the taboo are adaptations to the different ecologies of various parts of India (Harris, 1981).

Cultural materialism does not deny that ideas (beliefs) play an important role in the evolution of culture, but it does deny that they play a crucial role in initiating change. Consider the position of cultural materialism on the interesting question of why the practice of agriculture emerged when and where it did. Harris concedes that the practice itself requires certain underlying cognitive states--viz., agricultural knowledge. But the question is, why did these states arise? He maintains that they were not due to some rare combination of cognitive factors embodied in a long-forgotten genius. Instead, he asserts that human beings always had the potential for agriculture, but did not make use of this potential until more rewarding alternatives (such as hunting for big game) disappeared.

The type of explanation that Harris is proposing for the emergence of a novel cultural practice can be modeled by a thought experiment using concurrent schedules of reinforcement. Suppose one schedule is a variable-interval schedule and the other schedule is programmed to shape a novel response--say, turning around. The animal always has a choice between pecking the key or engaging in alternative responses. If the schedule for shaping is lean and the schedule for pecking is rich, then the animal will spend almost all its time at the key. But let the schedule on the key become lean enough, and the pigeon will eventually learn to turn.

In what sense do the contingencies of reinforcement explain the accomplishment of learning to turn? Ultimately, they explain it because they cause it to happen. When the returns on key pecking are good, no learning of a novel response occurs. When they are not, the process of learning a novel response begins. Of course the explanation of how the contingencies associated with shaping were themselves effective in producing a novel response might require a theory of underlying cognitive processes. But the answer to the question of why certain underlying cognitive capacities came to be exercised, even though they were there all along, would be behavioral. Hence, concurrent schedules can explain why certain capacities get put into use when and where they do.

This helps to explain the sense in which behavior analysis complements certain kinds of cognitive explanation and at the same time offers an alternative to strategies of therapeutic intervention based upon cognitive accounts. Consider the topic of self-control. Common sense conceives of self-control as some sort of inner agency that overwhelms the influence of the environment; but a commitment to environmental determinism requires a different analysis. In a classic paper, Rachlin (1974) argues that what self-control actually requires is a transfer of control over behavior from short-term environmental consequences to long-term environmental consequences. Suppose for example a person's social life is being disrupted by impulsive outbursts of anger which he wants to control but finds himself incapable of doing. He lacks, as we say, self-control. Rachlin views such cases as a problem of stimulus control. Righting the balance requires an intervention that shifts the balance from short-term consequences to long-term consequences.

Skinner (1953a) himself defined self-control as a special case of chaining. To chain two responses means to modify them in such a way that one response "may produce or alter some of the variables which control another response" (p. 224). The result is a sequence of responses in which the first member produces the second, the second produces the third, and so on. The rat's movement to the food magazine immediately after performing the response that delivers reinforcement is usually analyzed as a chain. The chain is initiated by the sound of the magazine operating, which causes the animal to turn in the direction of the magazine, which causes the animal to see the pellet, which causes the animal to approach the hopper, and so on, until the chain is terminated by the act of eating the pellet. Skinner writes that "a special kind of chaining is represented by *behavior which alters the strength of other behavior and is reinforced because it does so*" (p. 224, italics in original). He calls this type of behavior self-control, and says that it "could almost be said to distinguish the human organism from all others" (p. 224).

Such behavior alters the likelihood of other behavior and gets reinforced as a result of doing so. The reinforcement may come at the hands of other people, because they know that the behavior they are reinforcing alters other behavior in a manner they approve of. An example might be reinforcement of the public expression of remorse. Expressing remorse publicly decreases the probability of repeating the forbidden act. Such public displays of remorse get reinforced by the community because they reduce the probability of repeating the act. In other forms of self-control, however, the community is no longer necessary. The reinforcement of the response of remorse comes from the remorseful person himself, perhaps as the result of some mediating process such as avoidance. This is to say, by expressing remorse, the subject reduces the probability of repeating the forbidden act, and at the same time reduces the anxiety aroused by the inclination to perform the act. This reduction of anxiety is a form of negative reinforcement, i.e., it constitutes the removal of an aversive stimulus.

An organism capable of self-control has a repertoire of responses that are reinforced by the consequence of altering the probability of some other response. If turning away from an item one is tempted to purchase decreases the inclination to purchase it, and if this effect (decreasing the inclination to purchase the item) reinforces the act of turning away, then the act of turning away is an instance of self-control. Skinner analyzes many cultural practices as instances of self-control.

Counting to ten before speaking reduces the inclination to express one's anger. By giving someone the piece of advice to count to ten in order to control her anger (the advice itself being an instructional stimulus with function-altering effects), we cause the anger-reducing effect of counting to ten to function as a reinforcer for that action. The person's behavior may, as a result, come under the control of its long-term consequences (the improved social relations that result from inhibition of outbursts of anger) instead of the control of its short-term consequences (the distress suffered by the object of one's anger as a result of one's outburst). We thereby teach the subject to have self-control.

Another example of folk-wisdom is to advise dieters to put sweets out of reach. Doing so will reduce the inclination to reach for them (because it increases the effort required to get one's hands on them). If we advise someone who is on a diet to put the sweets on the top shelf of the least accessible cupboard, then this piece of advice may cause the inhibiting effects of the act of relocating the sweets to function as a reinforcer for that act. As a result, the person's behavior may come under the control of its long-term consequences (weight reduction) rather than its short term consequences (consumption of sweets)--i.e., there is an increase of self-control.

This process can also occur privately, and has led to interesting and effective forms of behavior therapy. Suppose a child's social life is being disrupted by frequent acts of hitting and punching. The therapist may discover that for this child a certain image (e.g., of a deep pool of water) reduces the inclination to hit and punch. The therapist then teaches the child to recognize private stimuli that precede aggressive acts. It is then possible to prompt the child to produce the image in a simulation of a circumstance which would normally increase the child's inclination to violence. Having prompted the child to produce the image and observed the child's nonviolent response, the therapist reinforces the child's act of producing the image (with praise or whatever else works for this particular child). The therapist does not know what inner stimuli precede the angry outbursts, but whatever they may be, they can eventually come to function discriminatively with respect to the imaging response. At this point the child produces the image without prompting from the therapist, but still receives a contrived form of reinforcement (from the therapist) for doing so. If treatment is successful, the contrived reinforcement gets replaced by the natural reinforcement that comes from controlling one's aggression (e.g., a higher level of reinforcement from one's peers). The imaging response is now part of a three-term contingency: an inner discriminative stimulus, an inner response, and social reinforcement. The long-term effects of the child's behavior (e.g., friendlier responses from other children) are now controlling operant responding rather than the short-term effects (e.g., inflicting harm on other children).[2]

The Relationship to Ethics. There are times when Skinner writes as if one could infer ethical principles directly from behavior analysis, on the grounds that ethical concepts such as good and bad are equivalent to basic behavioral concepts (Skinner, 1953a). Such direct inference from scientific theory to ethics would be uncharacteristic of modern science,[3] however, and I do not wish to defend such a point of view.

---

[2]Applications of the concept of self-control to the interpretation of everyday life can be found in Skinner (1953a). Kazdin (1978) provides experimental evidence of the effectiveness of rigorously defined self-control procedures for the purpose of modifying behavior. The use of inner actions and stimuli has been discussed at length by Meichenbaum (1977) under the heading of cognitive behavior modification.

[3]Humanistic psychology is a contemporary example of such an attempt to derive ethical conclusions from purely empirical premises. Carl Rogers, one of its leading figures, explicitly claimed that one could infer the human good from careful

I do, however, wish to explicate the relationship between behavior analysis and conclusions about ethics as exemplified in the actual behavior of behavior analysts.

In this respect, behavior analysts are typical modern scientists. They take certain values as given, and these define the proper goals of scientific applications. The contribution of science to ethical debate is to increase our knowledge of the means to these ends. Sometimes this knowledge changes our estimate of the costs and benefits of some practice. At first, what we knew about DDT was that it killed insect pests cheaply and efficiently. Our evaluation of it was positive. When we later discovered its effects on the reproductive cycle of bald eagles and peregrine falcons, we lost our enthusiasm for it.

When I observe behavior analysts deriving ethical conclusions from their theory, their inferences fit this pattern of means/ends rationality. A detailed knowledge of applied behavior analysis seems to impart a sense of proportion about the relative importance of various behavioral causes and effects, and this sense of proportion has ethical implications. Most behavior analysts, for example, believe that parents could make a significant improvement in their child rearing practices simply by replacing punishment with positive reinforcement.[4] They are not making the naive claim that this would solve all behavioral problems, but they think this strategy would reap significant benefits at a relatively small cost.

Indeed much of the success of applied behavior analysis is the result of teaching people how to put this maxim into practice. Donald Baer,

---

observation of human behavior. Rogers (1980) speaks of the direction people move in therapy. When a person is fully functioning--i.e., when certain obstacles to a positive evaluation of one's life are removed--she tends to make certain affirmations. We can take these affirmations to be scientific insights into the human good--subject, like any scientific inference, to correction on the basis of further evidence, but none the less substantiated scientifically.

[4]Positive reinforcement of a response occurs when the presence of the reinforcing stimulus increases the rate of responding. This contrasts with negative reinforcement, which occurs when the removal of a stimulus increases the rate of responding. Defined informally, positive reinforcement is the presentation of a pleasant stimulus, negative reinforcement is the removal of an unpleasant one.

one of the founders of applied behavior analysis, cites this as the principal reason why most applied behavior analysts are so slow to make use of the latest theoretical research.

> They have very little need to apply the newest basic findings. They have come upon an element of the old basic findings that for them is a revelation: the principle of positive reinforcement. A huge amount of the behavioral trouble that they can see in the world looks remarkably to them like the suddenly simple consequence of unapplied or misapplied positive reinforcement. If only they could get the missing contingencies going, or the misapplied ones shifted, they think that many of the problems at hand might be solved. The generality of that possibility is so apparent, and the difficulty of implementing just positive reinforcement in real-world terms is so formidable and so variable from problem situation to problem situation, that they have their hands full. (Baer, 1981, p. 88)

Obviously, this maxim is not a scientific principle, yet in some sense it rests upon scientific knowledge.

What do behavior analysts know that many of the rest of us do not? Their misgivings about punishment derive mainly from their awareness of its negative side effects. These include escape and avoidance, a narrowing of response repertoire, slowed learning, increased aggression, and even phobias and other behavioral disorders (Sidman, 1989). These costs of punishment do not, however, develop immediately, and therefore sometimes go unnoticed. And even when they are noticed, the benefits of punishment are immediate (the target behavior stops), but the costs are delayed. Behavioral psychologists know, however, that small immediate benefits can easily overwhelm large delayed costs. Thus, they are skeptical about the ability of an angry adult to weigh the costs and benefits of punishment accurately. Behavioral psychology does not prove that punishment is wrong, but after becoming steeped in the dynamics of behavior, people tend to arrive at the judgment that few people are qualified to use it beneficially. Furthermore, when behavior analysts have been permitted to put this judgment into clinical practice by helping people find non-coercive ways to create order, the net effect is usually judged good by the clients themselves.

I offer this example not to show that behavior analysis has reached a unique conclusion about the value of punishment (no doubt others have reached a similar conclusion without the benefit of an experimental program to guide them), but simply to illustrate the relationship between the type of knowledge behavior analysis aims at and a certain type of wisdom about behavior. Like most forms of wisdom, it is not easy to induce in others. Behavior analysts often speak of their frustration about knowing something with profound ethical implications, but not being able to convey these implications to others. Skinner has tried to do some of this in his popular books on behavioral psychology and morals, but his exposition becomes entangled in the doctrine of radical behaviorism. Sadly, his efforts probably set the cause back and made it more difficult than ever to make the case for positive reinforcement.

REFERENCES

Ainsley, G. (1985). Beyond microeconomics: Conflict among interests in a multiple self as a determinant of value. In J. Elster (Ed.), *The multiple self* (pp. 133-176). Cambridge: Cambridge University Press.

Alhadeff, D. A. (1982). *Microeconomics and human behavior: Towards a new synthesis of economics and psychology.* Berkeley: University of California Press.

Amundson, R. (1989). The trials and tribulations of selectionist explanations. In K. Halweg & C. A. Hooker (Eds.), *Issues in evolutionary epistemology* (pp. 413-432). Albany: State University of New York Press.

Andresen, J. (1991). Skinner and Chomsky 30 years later or: The return of the repressed. *The Behavior Analyst, 14,* 49-60.

Audi, R. (1976). B. F. Skinner on freedom, dignity, and the explanation of behavior. *Behaviorism, 4*, 163-186.

Baars, B. J. (1986). *The cognitive revolution in psychology.* New York: Guilford.

Baer, D. M. (1978). On the relation between basic and applied research. In A. C. Catania & T. A. Brigham (Eds.), *Handbook of applied behavior analysis* (pp. 9-16). New York: Irvington Publishers.

Baer, D. M. (1981). A flight of behavior analysis. *The Behavior Analyst, 4*, 85-91.

Bandura, A. (1974). Behavior theory and the models of man. *American Psychologist, 29*, 859-869.

Battalio, R. C. (1973). A test of consumer demand theory using observation of individual consumer purchases. *Western Economic Journal, 11*, 411-428.

Baum, W. (1974). On two types of deviation from the matching law: Bias and undermatching. *Journal of the Experimental Analysis of Behavior, 22*, 231-242.

Baum, W. M. (1989). Quantitative prediction and molar description of the environment. *The Behavior Analyst, 12*, 167-176.

Baxley, N., & Associates (Producer). (1982). *Cognition, creativity and behavior The Columban simulations* [Film]. Champaign, IL: Research Press.

Bechtel, W. (1984). Autonomous psychology: What it should and should not entail. *Proceedings of the 1984 biennial meeting of the Philosophy of Science Association* (Vol. 1, pp. 43-55). East Lansing, MI: Philosophy of Science Association.

Bechtel, W. (1988a). Perspectives on mental models. *Behaviorism, 16*, 137-148.

Bechtel, W. (1988b). *Philosophy of mind: An overview for cognitive science.* Hillsdale, NJ: Lawrence Erlbaum.

Bechtel, W. (1988c). *Philosophy of science: An overview for cognitive science.* Hillsdale, NJ: Lawrence Erlbaum.

Bechtel, W. & Abrahamsen, A. (1991). *Connectionism and the mind*. Cambridge, MA: Basil Blackwell.

Bechtel, W. & Richardson, R. C. (1993). *Discovering complexity: Decomposition and localization as strategies in scientific research*. Princeton: Princeton University Press.

Bolles, R. C. (1984). On the status of causal modes. *The Behavioral and Brain Sciences*, *7*, 482-483.

Bowler, P. J. (1983). *The eclipse of Darwinism: Anti-Darwinian evolution theories in the decades around 1900*. Baltimore: The Johns Hopkins University Press.

Bradshaw, C. M. & Szabadi, E. (1988). Quantitative analysis of human operant behavior. In G. Davey & C. Cullen (Eds.), *Human operant conditioning and behavior modification* (pp. 225-260). Chichester, England: John Wiley.

Brunswik, E. (1952). *The conceptual framework of psychology*. Chicago: University of Chicago Press.

Buskist, W. & Morgan, D. (1988). Method and theory in the study of human competition. In G. Davey & C. Cullen (Eds.), *Human operant conditioning and behavior modification* (pp. 167-196). Chichester, England: John Wiley.

Campbell, D. T. (1979). A tribal model of the social system vehicle carrying scientific knowledge. *Knowledge: Creation, Diffusion, Utilization*, *1*, 181-201.

Campbell, Donald T. (1987). Blind variation and selective retention in creative thought as in other knowledge processes. In G. Radnitzky and W. W. Bartley, III (Eds.), *Evolutionary epistemology, theory of rationality, and the sociology of knowledge* (pp. 91-114). LaSalle, IL: Open Court.

Catania, A. C. (1973a). The psychologies of structure, function, and development. *American Psychologist*, *28*, 434-443.

Catania, A. C. (1973b). Self-inhibiting effects of reinforcement. *Journal of the Experimental Analysis of Behavior*, *19*, 517-526.

Catania, A. C. (1976). Drug effects and concurrent performances. *Pharmacological Reviews*, *27*, 385-394.

Catania, A. C. (1979). *Learning*. Englewood Cliffs, NJ: Prentice-Hall, Inc.

Catania, A. C. (1981). Discussion: The flight from experimental analysis. In C. M. Bradshaw, E. Szabadi & C. F. Lowe (Eds.), *Quantification of steady-state behavior* (pp. 49-64). Elsevier, The Netherlands: North-Holland Biomedical Press.

Catania, A. C. (1987). Some Darwinian lessons for behavior analysis: A review of Bowler's *The eclipse of Darwinism*. *Journal of the Experimental Analysis of Behavior*, *47*, 249-257.

Catania, A. C. (1991). The gifts of culture and of eloquence: An open letter to Michael J. Mahoney in reply to his article, "Scientific psychology and radical behaviorism." *The Behavior Analyst*, *14*, 61-72.

Catania, A. C., & Harnad, S. (Eds.). (1984). The canonical papers of B. F. Skinner [Special Issue]. *Behavioral and Brain Sciences*, *7*(4).

Catania, A. C., & Harnad, S. (Eds.). (1988). *The selection of behavior. The operant behaviorism of B.F. Skinner: Comments and consequences*. Cambridge: Cambridge University Press.

Catania, A. C., Horne, P., & Lowe, C. F. (1989). Transfer of function across members of an equivalence class. *The Analysis of Verbal Behavior, 7,* 99-110.

Catania, A. C., Matthews, B. A., & Shimoff, E. (1982). Instructed versus shaped human verbal behavior: Interactions with nonverbal responding. *Journal of the Experimental Analysis of Behavior*, *38*, 233-248.

Catania, A. C., Shimoff, E., & Matthews, B. A. (1989). An experimental analysis of rule-governed behavior. In S. C. Hayes (Ed.), *Rule-governed behavior: Cognition, contingencies, and instructional control* (pp. 119-150). New York: Plenum Press.

Chomsky, N. (1956). Three models for the description of language. *IRE Transactions on Information Theory, IT-2,* 143-172.

Chomsky, N. (1959). Review of B. F. Skinner's *Verbal behavior*. *Language*, *35*, 26-58.

Chomsky, N. (1971, December). Review of B. F. Skinner's *Beyond freedom and dignity*. *New York Review of Books,* pp. 18-24.

Chomsky, N. (1975). *Reflections on language*. New York: Pantheon.

Chomsky, N. (1979). *Language and responsibility*. New York: Pantheon.

Chomsky, N. (1980). *Rules and representations*. New York: Columbia University Press.

Cohen, L. J. (1984). On the depth and fit of behaviorist explanation. *The Behavioral and Brain Sciences*, *7*, 591-592.

Coleman, S. R., & Mehlman, S. E. (1992). An empirical update (1969-1989) of D. L. Krantz's thesis that the experimental analysis of behavior is isolated. *The Behavior Analyst*, *15*, 43-49.

Commons, M. L., Kacelnik, A., & Shettleworth, S. J. (Eds.). (1987). *Quantitative analyses of behavior: Vol. 6. Foraging*. Hillsdale, NJ: Lawrence Erlbaum.

Cummins, R. (1983a). *The nature of psychological explanation*. Cambridge: Massachusetts Institute of Technology Press.

Cummins, R. (1983b). Analysis and subsumption in the behaviorism of Hull. *Philosophy of Science*, *50*, 96-111.

Davey, G. (1981). *Animal learning and conditioning*. Baltimore, MD: University Park Press.

Davidson, D. (1980). *Essays on actions and events*. New York: Oxford University Press.

Davidson, D. (1984). *Inquiries into truth and interpretation*. New York: Oxford University Press.

Davison, M. & McCarthy, D. (1988). *The matching law: A research review*. Hillsdale, NJ: Lawrence Erlbaum.

Day, W. F. (1983). On the difference between radical and methodological behaviorism. *Behaviorism, 11*, 89-102.

Day, W. F. (1987). What is radical behaviorism? In S. Modgil & C. Modgil (Eds.), *B. F. Skinner: Consensus and controversy* (pp. 13-40). New York: Falmer Press.

Deitz, S. M. (1978). Current status of applied behavior analysis: Science versus technology. *American Psychologist, 33*, 805-814.

Dennett, D. (1978). *Brainstorms.* Cambridge: Massachusetts Institute of Technology Press.

Dennett, D. (1984). *Elbow room: The varieties of free will worth wanting.* Cambridge: Massachusetts Institute of Technology Press.

de Villiers, P. A. (1977). Choice in concurrent schedules and a quantitative formulation of the law of effect. In W. K. Honig & J. E. R. Staddon (Eds.), *Handbook of operant behavior* (pp. 233-287). Englewood Cliffs, NJ: Prentice-Hall.

Dretske, F. I. (1969). *Seeing and knowing.* Chicago: University of Chicago Press.

Dretske, F. I. (1981). *Knowledge and the flow of information.* Cambridge Massachusetts Institute of Technology Press.

Dretske, F. I. (1988). *Explaining behavior: Reasons in a world of causes.* Cambridge: Massachusetts Institute of Technology Press.

Dretske, F. I. (1989). Reasons and causes. In J. E. Tomberlin (Ed.), *Philosophical perspectives*, (Vol. 3, pp. 1-16). Atascadero, CA: Ridgeview Publishing Company.

Elster, J. (1979). *Ulysses and the sirens: Studies in rationality and irrationality.* Cambridge: Cambridge University Press.

Enc, B. & Adams, F. (1992). Functions and goal directedness. *Philosophy of Science, 59*, 635-654.

Epstein, R. (1981). On pigeons and people: A preliminary look at the Columban Simulation Project. *The Behavior Analyst, 4*, 43-55.

Epstein, R. (1984). The case for praxics. *The Behavior Analyst, 7*, 47-59.

Estes, W. K., Koch, S., MacCorquodale, K., Meehl, P. E., Mueller, C. G., Schoenfeld, W. N., & Verplanck, W. S. (1954). *Modern learning theory: A critical analysis of five examples.* New York: Appleton-Century-Crofts.

Falk, J. L. (1986). The formation and function of ritual behavior. In T. Thompson & M. D. Zeiler (Eds.), *Analysis and integration of behavioral units* (pp. 335-355). Hillsdale, NJ: Lawrence Erlbaum.

Ferster, C. B. & Perrott, M. C. (1968). *Behavior principles.* New York: Appleton-Century-Crofts.

Ferster, C. B., & Skinner, B. F. (1957). *Schedules of reinforcement.* New York: Appleton-Century-Crofts.

Fetzer, James H. (1990). *Artificial intelligence: Its scope and limits.* Dordrecht, The Netherlands: Kluwer.

Findley, J. D. (1958). Preference and switching under concurrent scheduling. *Journal of the Experimental Analysis of Behavior*, *1*, 123-144.

Fodor, J. A. (1975). *The language of thought*. New York: Thomas Y. Crowell Company.

Fodor, J. A. (1980). Methodological solipsism considered as a research strategy in cognitive psychology. *The Behavioral and Brain Sciences*, *3*, 417-424.

Fodor, J. A. (1981, January). The mind-body problem. *The Scientific American*, pp. 114-123. Reprinted in J. Feinberg (Ed.), *Reason & responsibility* (7th ed., pp. 287-297). Belmont, CA: Wadsworth Publishing.

Fodor, J. A. (1983). *The modularity of mind*. Cambridge: Massachusetts Institute of Technology Press.

Fraley, L. E. & Vargas, E. A. (1986). Separate disciplines: The study of behavior and the study of the psyche. *The Behavior Analyst*, *9*, 47-59.

Fuqua, R. W. (1984). Comments on the applied relevance of the matching law. *Journal of Applied Behavior Analysis*, *17*, 381-386.

Garcia, J., Kimmeldorf, D. J., & Hunt, E. L. (1961). The use of ionizing radiation as a motivating stimulus. *Psychological Review*, *68*, 383-385.

Garcia, J., & Koelling, R. A. (1966). The relation of cue to consequence in avoidance learning. *Psychonomic Science*, *4*, 123-124.

Gautier, D. (1983). Review of Jon Elster's *Ulysses and the Sirens: Studies in rationality and irrationality*. *Canadian Journal of Philosophy*, *13*, 133-140.

Gewirtz, J. L., & Pelaez-Nogueras, M. (1992). B. F. Skinner's legacy to human infant behavior and development. *American Psychologist*, *47* (11), 1411-1422.

Gholson, B. & Barker, P. (1985). Kuhn, Lakatos, and Laudan: Applications in the history of physics and psychology. *American Psychologist*, *40*, 755- 769.

Gilgen, A. R. (1982). *American psychology since World War II: A profile of the discipline*. Westport, CT: Greenwood Press.

Goodman, N. (1955). *Fact, fiction and forecast*. Cambridge: Harvard University Press.

Graham, G. (1977). On what is good: A study of B. F. Skinner's operant behaviorist view. *Behaviorism*, *5*, 97-112.

Graham, G. (1983). More on the goodness of Skinner. *Behaviorism*, *11*, 45-51.

Guttman, N. (1977). On Skinner and Hull: A reminiscence and projection. *American Psychologist*, *32*, 321-328.

Hammond, L. J. (1980). The effect of contingency upon the appetitive conditioning of free operant behavior. *Journal of the Experimental Analysis of Behavior*, *34*, 297-304.

Harris, M. (1966). Monistic determinism: Anti-service. *Southwestern Journal of Anthropology*, *25*, 198-205.

Harris, M. (1974). *Cows, pigs, wars and witches: The riddles of cultures*. New York: Random House.

Harris, M. (1977). *Cannibals and kings: The origins of cultures*. New York: Random House.

Harris, M. (1981). Sacred cows and water buffalo in India: The uses of ethnography. *Current Anthropology*, *22*, 483-489.

Hayes, L. J., & Chase, P. N. (Eds.). (1991). *Dialogues on verbal behavior*. Reno, NV: Context Press.

Hayes, S. D. (1989). Non-humans have not yet shown stimulus equivalence. *Journal of the Experimental Analysis of Behavior*, *51*, 585-592.

Hayes, S. C., Kohlenberg, B. S., & Melancon, S. M. (1989). Avoiding and altering rule-control as a strategy of clinical intervention. In S. C. Hayes (Ed.), *Rule-governed behavior: Cognition, contingencies, and instructional control* (pp. 359-386). New York: Plenum Press.

Herrnstein, R. J. (1961). Relative and absolute strength of response as a function of frequency of reinforcement. *Journal of the Experimental Analysis of Behavior*, *4*, 267-272.

Herrnstein, R. J. (1970). On the law of effect. *Journal of the Experimental Analysis of Behavior*, *13*, 243-266.

Herrnstein, R. J. (1990). Rational choice theory: Necessary but not sufficient. *American Psychologist*, *45*, 356-367.

Herrnstein, R. J. (1991). Reply to Binmore and Staddon. *American Psychologist*, *46*, 799-801.

Herrnstein, R. J. and Hineline, P. N. (1966). Negative reinforcement as shock-frequency reduction. *Journal of the Experimental Analysis of Behavior*, *9*, 421-430.

Herrnstein, R. J., Loveland, D. H., & Cable, C. (1976). Natural concepts in pigeons. *Journal of Experimental Psychology: Animal Behavior Processes*, *2*, 285-302.

Herrnstein, R. J. & Vaughan, W., Jr. (1980). Melioration and behavioral allocation. In J. E. R. Staddon (Ed.), *Limits to action: The allocation of individual behavior* (pp. 143-176). New York: Academic Press.

Hilgard, E. R. (1956). *Theories of learning* (2nd ed.). New York: Appleton-Century-Crofts.

Hilgard, E. R. and Bower, G. H. (1975). *Theories of learning* (4th ed.). Englewood Cliffs, NJ: Prentice-Hall.

Hineline, P. N. (1984). Editorial. *Journal of the Experimental Analysis of Behavior*, *41*, 1-2.

Hinson, J. (1987). Skinner and the unit of behavior. In S. Modgil & C. Modgil (Eds.), *B. F. Skinner: Consensus and controversy* (pp. 181-192). New York: Falmer Press.

Hull, C. L. (1943). *Principles of behavior*. New York: Appleton-Century-Crofts.

Hursh, S. R. (1980). Economic concepts for the analysis of behavior. *Journal of the Experimental Analysis of Behavior*, *34*, 219-238.

Hursh, S. R. (1984). Behavioral economics. *Journal of the Experimental Analysis of Behavior*, *42*, 435-452.

Hyten, C., & Reilly, M. P. (1992). The renaissance of the experimental analysis of human behavior. *The Behavior Analyst*, *15*, 109-114.

Katahn, M., & Koplin, J. H. (1968). Paradigm clash: Comment on "Some recent criticisms of behaviorism and learning theory with special reference to Breger and McGaugh and to Chomsky." *Psychological Bulletin*, *69*, 147-148.

Kaufman, A., Baron, A., & Kopp, R. E. (1966). Some effects of instructions on human operant behavior. *Psychonomic Monograph Supplements*, *11*, 243-350.

Kazdin, A. E. (1978). *History of behavior modification: Experimental foundations of contemporary research*. Baltimore, MD: University Park Press.

Killeen, P. F. (1987). Emergent behaviorism. In S. Modgil & C. Modgil (Eds.), *B. F. Skinner: Consensus and controversy* (pp. 219-234). Philadelphia, PA: Falmer Press.

Kim, J. (1984). Supervenience and supervenient causation. In T. Horgan (Ed.), *Spindel Conference 1983: Supervenience.* Supplement to *The Southern Journal of Philosophy*, *22*, 45-56.

Kitcher, P. (1989). Proximate and developmental analysis. *The Behavioral and Brain Sciences*, *12*, 186-187.

Koch, S. (1964). Psychology and emerging conceptions of knowledge as unitary. In T. W. Wann (Ed.), *Behaviorism and phenomenology: Contrasting bases for modern psychology* (pp. 1-41). Chicago: University of Chicago Press.

Krantz, D. L. (1971). The separate worlds of operant and non-operant psychology. *Journal of Applied Behavior Analysis*, *4*, 61-70.

Krantz, D. L. (1972). Schools and systems: The mutual isolation of operant and non-operant psychology as a case study. *Journal of the History of the Behavioral Sciences*, *8*, 86-102.

Krantz, D. L. & Wiggins, L. (1973). Personal and impersonal channels of recruitment in the growth of theory. *Human Development*, *16*, 133-156.

Krebs, J. R., & Davies, N. B. (Eds.). (1978). *Behavioral ecology*. Oxford, England: Blackwell Scientific Publications.

Kuhn, T. S. (1962). *The structure of scientific revolutions*. Chicago: University of Chicago Press.

Kuhn, T. S. (1970). *The structure of scientific revolutions* (2nd ed.). Chicago: University of Chicago Press.

Kuhn, T. S. (1992). *The trouble with the historical philosophy of science*. Cambridge: Department of the History of Science, Harvard University.

Lacey, H. (1979). Skinner on the prediction and control of behavior. *Theory and Decision*, *10*, 353-385.

Lacey, H., & Rachlin, H (1978). Behavior, cognition and theories of choice. *Behaviorism*, *6*, 177-202.

Lacey, H. & Schwartz, B. (1986). Behaviorism, intentionality, and socio-historical structure. *Behaviorism*, *14*, 193-210.

Lacey, H. & Schwartz, B. (1987). The explanatory power of radical behaviorism. In S. Modgil and C. Modgil (Eds.), *B. F. Skinner: Consensus and controversy* (pp. 165-176). New York: Falmer Press.

Lachman, R., Lachman, J., & Butterfield, E. (1979). *Cognitive psychology and information processing*. Hillsdale, NJ: Lawrence Erlbaum.

Lashley, K. (1951). The problem of serial order in behavior. In L. A. Jeffress (Ed.), *Cerebral mechanisms in behavior: The Hixon symposium* (pp. 112-136). New York: John Wiley.

Lattal, K. A. (Ed.). (1992). Reflections on B. F. Skinner and psychology [Special Issue]. *American Psychologist*, *47*(11).

Lattal, K. A. & Harzem, P. (1984). Present trends and directions for the future. *Journal of the Experimental Analysis of Behavior*, *42*, 349-352.

Lea, S. E. G. (1981). Correlation and contiguity in foraging theory. In P. Harzem & M. D. Zeiler (Eds.), *Advances in analysis of behavior: Vol. 2. Predictability, correlation, and contiguity* (pp. 355-406). New York: John Wiley.

Leahey, T. H. (1980). *A history of psychology: Main currents in psychological thought*. Englewood Cliffs, NJ: Prentice-Hall.

Leahey, T. H. (1987). *A history of psychology: Main currents in psychological thought* (2nd ed.). Englewood Cliffs, NJ: Prentice-Hall.

Leahey, T. H. (1992). The mythical revolutions of American psychology. *American Psychologist*, *47*, 308-318.

Lee, V. (1988). *Beyond behaviorism*. Hillsdale, NJ: Lawrence Erlbaum.

Lloyd, K. E. (1985). Behavioral anthropology: A review of Marvin Harris' *Cultural materialism*. *Journal of the Experimental Analysis of Behavior*, *43*, 279-287.

Lorenz, K. (1970). Taxis and instinctive behavior pattern in egg-rolling by the Greylag goose. In K. Lorenz, *Studies in animal and human behavior* (Vol. 1, pp. 316-350). Cambridge: Harvard University Press.

Losee, J. (1980). *A historical introduction to the philosophy of science* (2nd ed.). Oxford: Oxford University Press.

Lowe, C. F. (1979). Determinants of human operant behavior. In M.D. Zeiler & P. Harzem (Eds.) *Advances in the analysis of behavior: Vol. 1. Reinforcement and the organization of behavior* (pp. 159-192). New York: John Wiley.

Lowe, C. F., & Horne, P. J. (1985). On the generality of behavioral principles: Human choice and the matching law. In C. F. Lowe, M. Richelle, D. E. Blackman & C. M. Bradshaw (Eds.), *Behavior analysis & contemporary psychology* (pp. 97-116). London: Lawrence Erlbaum.

MacCorquodale, K. (1970). On Chomsky's review of Skinner's *Verbal behavior*. *Journal of the Experimental Analysis of Behavior*, *13*, 83-99.

MacCorquodale, K. & Meehl, P. E. (1948). On a distinction between hypothetical constructs and intervening variables. *Psychological Review*, *55*, 95-107.

Mackenzie, B. D. (1977). *Behaviorism and the limits of scientific method.* Atlantic Highlands, NJ: Humanities Press.

Mahoney, M. J. (1989). Scientific psychology and radical behaviorism: Important distinctions based in scientism and objectivism. *American Psychologist*, *44*, 1372-1377.

Malone, J. C. (1987a). Skinner, the behavioral unit, and current psychology. In S. Modgil & C. Modgil (Eds.), *B. F. Skinner: Consensus and controversy* (pp. 193-203). New York: Falmer Press.

Malone, J. C. (1987b). Malone replies to Hinson. In S. Modgil & C. Modgil (Eds.), *B. F. Skinner: Consensus and controversy* (pp. 205-206). New York: Falmer Press.

Maltzman, I. R. (1986). A spirited defense and criticism. In B. J. Baars, *The cognitive revolution in psychology* (pp. 99-109). New York: Guilford Press.

Margolis, J. (1984). *Philosophy of psychology*. Englewood Cliffs, NJ: Prentice-Hall.

Marr, M. J. (1984). Conceptual approaches and issues. *Journal of the Experimental Analysis of Behavior*. *42*, 353-362.

Marr, M. J. (1989). Some remarks on the quantitative analysis of behavior. *The Behavior Analyst*, *12*, 143-152.

Matthews, B. A., Shimoff, E., Catania, A. C., & Sagvolden, T. (1977). Uninstructed human responding: Sensitivity to ratio and interval contingencies. *Journal of the Experimental Analysis of Behavior*, *27*, 453-467.

McDowell, J. J. (1982). The importance of Herrnstein's mathematical statement of the law of effect for behavior therapy. *American Psychologist*, *37*, 771-779.

McDowell, J. J. (1988). Matching theory in natural human environments. *The Behavior Analyst*, *11*, 95-109.

McDowell, J. J. (1989). Two modern developments in matching theory. *The Behavior Analyst*, *12*, 153-166.

Meehl, P. (1950). On the circularity of the law of effect. *Psychological Bulletin*, *47*, 52-75.

Meehl, P. (1986). Trait language and behaviorese. In T. Thompson & M. Zeiler (Eds.), *Analysis and integration of behavioral units* (pp. 315-334). London: Lawrence Erlbaum.

Meichenbaum, D. (1977). *Cognitive behavior modification*. New York: Plenum Press.

Michael, J. (1982). Distinguishing between discriminative and motivational functions of stimuli. *Journal of the Experimental Analysis of Behavior*, *37*, 149-155.

Michael, J. (1984). Verbal behavior. *Journal of the Experimental Analysis of Behavior*, *42*, 363-376.

Michael, J. (1985). Fundamental research and behavior modification. In C. F. Lowe, M. Richelle, D. E. Blackman & C. M. Bradshaw (Eds.), *Behavior analysis & contemporary psychology* (pp. 159-164). London: Lawrence Erlbaum.

Mill, J. S. (1865). *System of logic*. London: Longmans, Green.

Miller, G. A. & Chomsky, N. (1963). Finitary models of language users. In R. D. Luce, R. R. Bush, & E. Galanter (Eds.), *Handbook of mathematical psychology* (Vol. 2, pp. 419-492). New York: John Wiley.

Morris, C. (1958). Review of B. F. Skinner's *Verbal behavior*. *Contemporary Psychology*, *3*, 212-214.

Morse, W. H. & Kelleher, R. T. (1977). Determinants of reinforcement and punishment. In W. Honig, & J. E. R. Staddon (Eds.), *Handbook of operant behavior* (pp. 174-200). Englewood Cliffs, NJ: Prentice-Hall.

Myerson, J. & Hale, S. (1984). Practical implications of the matching law. *Journal of Applied Behavior Analysis*, *17*, 367-380.

Neisser, U. (1972). A paradigm shift in psychology: A review of A. Richardson's *Mental imagery*; S. J. Segal's (Ed.), *Imagery: Current cognitive approaches*; and A. Paivio's *Imagery and verbal processes*. *Science*, *176*, 628-630.

Nelson, R. J. (1969). Behaviorism is false. *Journal of Philosophy*, *66*, 417-451.

Nelson, R. J. (1982). *The logic of mind*. Dordrecht, The Netherlands: D. Reidel Publishing Company.

Nelson, R. J. (1984). Skinner's philosophy of method. *The Behavioral and Brain Sciences*, *7*, 529-530.

Nelson, R. J. (1989). Review of G. E. Zuriff's *Behaviorism: A conceptual reconstruction*. *Synthese*, *80*, 305-313.

Nevin, J. A. (1980). Editorial. *Journal of the Experimental Analysis of Behavior*, *34*, 133-134.

Nevin, J. A. (1984). Quantitative analysis. *Journal of the Experimental Analysis of Behavior*, *42*, 421-434.

Newmeyer, F. I. (1980). *Linguistic theory in America: The first quarter century of transformational generative grammar*. New York: Academic Press.

Osgood, C. E. (1958). Review of B. F. Skinner's *Verbal behavior*. *Contemporary Psychology*, *3*, 209-212.

Pierce, W. D., & Epling, W. F. (1980). What happened to analysis in applied behavior analysis? *The Behavior Analyst*, *1*, 1-9.

Pierce, W. D., & Epling, W. F. (1991). Can operant research with animals rescue the science of human behavior? *The Behavior Analyst*, *14*, 129-132.

Pitts, R. C. & Malagodi, E. F. (1991). Preference for less frequent shock under fixed-interval schedules of electric-shock presentation. *Journal of the Experimental Analysis of Behavior*, *56*, 21-32.

Premack, D. (1965). Reinforcement theory. In D. Levine (Ed.), *Nebraska symposium on motivation* (Vol. 13). Lincoln: University of Nebraska Press.

Proctor, R. W. & Weeks, D. J. (1990). *The goal of B. F. Skinner and behavior analysis*. New York: Springer-Verlag.

Putnam, H. (1975). The nature of mental states. In *Mind, language, and reality* (pp. 139-152). Cambridge: Cambridge University Press.

Quine, W. V. O. (1953). *From a logical point of view*. Cambridge: Harvard University Press.

Rachlin, H. (1974). Self-control. *Behaviorism, 2*, 94-107.
Rachlin, H. (1976). *Modern behaviorism.* San Francisco: W. H. Freeman.
Rachlin, H. (1978). A molar theory of reinforcement schedules. *Journal of the Experimental Analysis of Behavior, 30*, 345-360.
Rachlin, H. (1980). Economics and behavioral psychology. In J.E.R. Staddon (Ed.), *Limits to action.* New York: Academic Press.
Rachlin, H. (1989). *Judgment, decision and choice: A cognitive/behavioral synthesis.* New York: W. H. Freeman.
Rachlin, H. (1992). Teleological behaviorism. *American Psychologist, 47*, 1371-1382.
Rachlin, H., & Green, L. (1972). Commitment, choice and self-control. *Journal of the Experimental Analysis of Behavior, 17*, 15-22.
Rachlin, H., Green, L., Kagel, J. H., & Battalio, R. C. (1976). Economic demand theory and psychological theories of choice. In G. H. Bower (Ed.), *The psychology of learning and motivation* (Vol. 10, pp. 129-154). New York: Academic Press.
Rescorla, R. A. (1967). Pavlovian conditioning and its proper control procedures. *Psychological Review, 74*, 71-80.
Rescorla, R. A. (1975). Pavlovian excitatory and inhibitory conditioning. In W. K. Estes (Ed.), *Handbook of learning and cognitive processes: Vol. 2. Conditioning and behavior theory* (pp. 5-35). Hillsdale, NJ: Lawrence Erlbaum.
Rice, B. (1968, March 17). B. F. Skinner: The most important influence on modern psychology. *The New York Times Magazine*, pp. 127-137.
Ringen, J. D. (1976). Explanation, teleology, and operant behaviorism: A study of the experimental analysis of purposive behavior. *Philosophy of Science, 43*, 223-253.
Ringen, J. D. (1985). Operant conditioning and a paradox of teleology. *Philosophy of Science, 52*, 565-577.
Ringen, J. D. (1986). The completeness of behavior theory: A review of Barry Schwartz and Hugh Lacey's *Behaviorism, science, and human nature. Behaviorism, 14*, 29-40.
Rogers, C. R. (1980). *A way of being*. Boston: Houghton Mifflin Company.
Rosenberg, A. (1978). The supervenience of biological concepts. *Philosophy of Science, 45*, 368-386.
Rosenberg, A. (1988). *Philosophy of social science*. Boulder, CO: Westview Press.
Schlinger, H. & Blakely, E. (1987). Function-altering effects of contingency-specifying stimuli. *The Behavior Analyst, 10*, 41-45.
Schnaitter, R. (1984). "Behaviorism at fifty" at twenty. *The Behavioral and Brain Sciences, 7*, 644-645.
Schnaitter, R. (1987). Behaviorism is not cognitive and cognitivism is not mental. *Behaviorism, 15*, 1-11.

Schwartz, B. (1986). *The battle for human nature: Science, morality and modern life*. New York: Norton.

Schwartz, B. & Lacey, H. (1982). *Behaviorism, science, and human nature*. New York: Norton.

Schwartz, B. & Lacey, H. (1988). What applied studies of human operant conditioning tell us about humans and about operant conditioning. In G. Davey & C. Cullen (Eds.), *Human operant conditioning and behavior modification* (pp. 27-42). New York: John Wiley.

Scriven, M. (1956). A study of radical behaviorism. In H. Feigl & M. Scriven (Eds.), *Minnesota studies in the philosophy of science* (Vol. 1, pp. 88-130). Minneapolis: University of Minnesota Press.

Segal, E. F. (1977). Toward a coherent psychology of language. In W. K. Honig & J. E. R. Staddon (Eds.), *Handbook of operant behavior* (pp. 628-653). Englewood Cliffs, NJ: Prentice-Hall.

Segal, E. M., & Lachman, R. (1972). Complex behavior or higher mental process: Is there a paradigm shift? *American Psychologist*, *27*, 46-55.

Shimp, C. P. (1966). Probabilistically reinforced choice behavior in pigeons. *Journal of the Experimental Analysis of Behavior*, *9*, 443-455.

Shimp, C. P. (1984). Cognition, behavior, and the experimental analysis of behavior. *Journal of the Experimental Analysis of Behavior*, *42*, 407-420.

Sidman, M. (1960). *Tactics of scientific research*. New York: Basic Books.

Sidman, M. (1971). Reading and auditory-visual equivalences. *Journal of Speech and Hearing Research*, *14*, 5-13.

Sidman, M. (1979). Remarks. *Behaviorism*, *7*, 123-126.

Sidman, M. (1986). Functional analysis of emergent verbal classes. In T. Thompson & M. D. Zeiler (Eds.), *Analysis and integration of behavioral units* (pp. 213-245). Hillsdale, NJ: Lawrence Erlbaum.

Sidman, M. (1989). *Coercion and its fallout*. Boston: Authors Cooperative.

Sidman, M., Rauzin, R., Lazar, R., Cunningham, S., Tailby, W., & Carrigan, P. (1982). A search for symmetry in the conditional discriminations of rhesus monkeys, baboons, and children. *Journal of the Experimental Analysis of Behavior*, *37*, 23-44.

Skinner, B. F. (1931). The concept of the reflex in the description of behavior. *The Journal of General Psychology*, *5*, 427-458. Reprinted in B. F. Skinner (1961), *Cumulative record* (pp. 319-346). New York: Appleton-Century-Crofts.

Skinner, B. F. (1935a). The generic nature of the concepts of stimulus and response. *The Journal of General Psychology*, *12*, 40-65. Reprinted in B. F. Skinner (1961), *Cumulative record* (pp. 347-366). New York: Appleton-Century-Crofts.

Skinner, B. F. (1935b). Two types of conditioned reflex and a pseudo-type. *The Journal of General Psychology, 12*, 66-77. Reprinted in B. F. Skinner (1961), *Cumulative record* (pp. 367-376). New York: Appleton-Century-Crofts.

Skinner, B. F. (1937). Two types of conditioned reflex: A reply to Konorski and Miller. *The Journal of General Psychology, 16*, 264-272. Reprinted in B. F. Skinner (1961), *Cumulative record* (pp. 376-383). New York: Appleton-Century-Crofts.

Skinner, B. F. (1938). *The behavior of organisms: An experimental analysis*. New York: Appleton-Century-Crofts.

Skinner, B. F. (1945). The operational analysis of psychological terms. *Psychological Review, 52*, 270-277;291-294. Reprinted in Skinner, B. F. (1961), *Cumulative record*, Enlarged edition (pp. 272-286). New York: Appleton-Century-Crofts.

Skinner, B. F. (1948a). 'Superstition' in the pigeon. *Journal of Experimental Psychology, 38*, 168-172.

Skinner, B. F. (1948b). *Walden two*. New York: Macmillan.

Skinner, B. F. (1950). Are theories of learning necessary? *Psychological Review, 57*, 193-216. Reprinted in B. F. Skinner (1961), *Cumulative record* (pp. 39-69). New York: Appleton-Century-Crofts.

Skinner, B. F. (1953a). *Science and human behavior*. New York: The Free Press.

Skinner, B. F. (1953b). Some contributions of an experimental analysis of behavior to psychology as a whole. *American Psychologist, 8*, 69-78.

Skinner, B. F. (1956). A case history in scientific method. *American Psychologist, 11*, 221-233. Reprinted in B. F. Skinner (1961), *Cumulative record* (pp. 76-99). New York: Appleton-Century-Crofts.

Skinner, B. F. (1957). *Verbal behavior*. New York: Appleton-Century-Crofts.

Skinner, B. F. (1963a). Behaviorism at fifty. *Science, 140*, 951-958. Reprinted with commentary in B. F. Skinner (1969), *Contingencies of reinforcement: A theoretical analysis* (pp. 221-268). Englewood Cliffs, NJ: Prentice-Hall.

Skinner, B. F. (1963b). Operant behavior. *American Psychologist, 18*, 503-515. Reprinted with commentary in B. F. Skinner (1969), *Contingencies of reinforcement: A theoretical analysis* (pp. 105-132). Englewood Cliffs, NJ: Prentice-Hall.

Skinner, B. F. (1966a). The phylogeny and ontogeny of behavior. *Science, 153*, 1205-1213. Reprinted with commentary in B. F. Skinner (1969), *Contingencies of reinforcement: A theoretical analysis* (pp. 172-220). Englewood Cliffs, NJ: Prentice-Hall.

Skinner, B. F. (1966b). An operant analysis of problem solving. In B. Kleinmuntz (Ed.), *Problem solving: Research, method, and theory* (pp 225-257). New York: John Wiley. Reprinted with commentary in B. F. Skinner (1969), *Contingencies of reinforcement: A theoretical analysis* (pp. 133-171). Englewood Cliffs, NJ: Prentice-Hall.

Skinner, B. F. (1967). B. F. Skinner. In E. G. Boring & G. Lindzey (Eds.), *A history of psychology in autobiography* (Vol. 5, pp. 387-413). New York: Appleton-Century-Crofts.

Skinner, B. F. (1969). *Contingencies of reinforcement: A theoretical analysis*. Englewood Cliffs, NJ: Prentice-Hall.

Skinner, B. F. (1971). *Beyond freedom and dignity*. New York: Bantam/Vintage Book, Random House.

Skinner, B. F. (1972). John Broadus Watson, behaviorist. In B. F. Skinner, Cumulative record (2nd ed., pp. 555-558). New York: Appleton-Century-Crofts.

Skinner, B. F. (1974). *About behaviorism*. New York: Alfred A. Knopf.

Skinner, B. F. (1979). *The shaping of a behaviorist*. New York: Alfred A. Knopf.

Skinner, B. F. (1980). *Notebooks: B. F. Skinner*. Edited by R. Epstein. Englewood Cliffs, NJ: Prentice-Hall.

Skinner, B. F. (1981). Selection by consequences. *Science*, *213*, 501-504.

Skinner, B. F. (1983a). *A matter of consequences*. New York: Alfred A. Knopf.

Skinner, B. F. (1983b). Taking our future into our hands. In J. E. Morrow (Chair), *Changing the course: Behaviorism's role in preventing global disaster*. Symposium conducted at the 9th annual convention of the Association for Behavior Analysis, Milwaukee, Wisconsin.

Skinner, B. F. (1984a). *A matter of consequences*. New York: New York University Press.

Skinner, B. F. (1984b). Author's response. *The Behavioral and Brain Sciences*, *7*, 655-667.

Skinner, B. F. (1987). *Upon further reflection*. Englewood Cliffs, NJ: Prentice-Hall.

Skinner, B. F. (1990). Can psychology be a science of mind? *American Psychologist*, *45*, 1206-1210.

Smart, J. J. C. (1961). Free will, praise, and blame. *Mind*, *70*, 291-306.

Smith, L. D. (1986). *Behaviorism and logical positivism: A reassessment of the alliance*. Stanford: Stanford University Press.

Smith, L. D. (1992). On prediction and control: B. F. Skinner and the technological ideal of science. *American Psychologist*, *47*, 216-223.

Smith, T. L. (1982). The all-and-only idealization. In K. Tuite, R. Schneider & R. Chametzky (Eds.), *Papers from the eighteenth regional meeting of the Chicago Linguistic Society* (pp. 477-487). Chicago: Chicago Linguistic Society.

Smith, T. L. (1983). Skinner's environmentalism: The analogy with natural selection. *Behaviorism*, *11*, 133-153.

Smith, T. L. (1984). The complex case of Brownstein and Shull's review of Schwartz and Lacey. *The Behavior Analyst*, *7*, 213-214.

Smith, T. L. (1986). Biology as allegory: A review of Elliott Sober's *The nature of selection*. *Journal of the Experimental Analysis of Behavior*, *46*, 105-112.

Smith, T. L. (1988). Neo-Skinnerian psychology: A non-radical behaviorism. *Proceedings of the 1988 biennial meeting of the Philosophy of Science Association*, (Vol. 1, pp. 143-148). East Lansing, MI: Philosophy of Science Association.

Smith, T. L. (1991). *A neoskinnerian analysis of operant psychology: Prospects for a non-radical behaviorism*. (Grant No. DIR-89121291). Washington, D.C.: National Science Foundation.

Sober, E. (1983). Mentalism and behaviorism in comparative psychology. In D.W. Rajecki (Ed.), *Comparing behavior: Studying man studying animals* (pp. 113-142). Hillsdale, NJ: Lawrence Erlbaum.

Sober, E. (1984). *The nature of selection: Evolutionary theory in philosophical focus*. Cambridge: Massachusetts Institute of Technology Press.

Sober, E. (1985a). Methodological behaviorism, evolution, and game theory. In J. H. Fetzer (Ed.), *Sociobiology and epistemology* (pp. 181-200). Dordrecht, The Netherlands: D. Reidel Publishing.

Sober, E. (1985b). Review of A. Rosenberg's *Sociobiology and the preemption of social science*. *Philosophy of the Social Sciences*, *15*, 89-93.

Sober, E. (1987). What is adaptationism? In J. Dupre (Ed.), *The latest on the best: Essays on evolution and optimality* (pp. 105-118). Cambridge: Massachusetts Institute of Technology Press.

Sosa, E. (1984). Behavior, theories, and the inner. *The Behavioral and Brain Sciences*, *7*, 537-539.

Staddon, J. E. R. (1967). Asymptotic behavior: The concept of the operant. *Psychological Review*, *74*, 377-391.

Staddon, J. E. R. (1973). On the notion of cause with application to behaviorism. *Behaviorism*, *2*, 25-63.

Staddon, J. E. R. (1975). Learning as adaptation. In W. K. Estes (Ed.), *Handbook of learning and cognitive processes* (pp. 37-98). Hillsdale, NJ: Lawrence Erlbaum.

Staddon, J. E. R. (Ed.). (1980). *Limits to action: The allocation of individual behavior*. New York: Academic Press.

Staddon, J. E. R. (1983). *Adaptive behavior and learning*. Cambridge: Cambridge University Press.

Staddon, J. E. R. (1984). Social learning theory and the dynamics of interaction. *Psychological Review*, *91*, 502-507.

Staddon, J. E. R. (1987). Optimality theory and behavior. In J. Dupre (Ed.), *The latest on the best* (pp. 179-198). Cambridge: Massachusetts Institute of Technology Press.

Staddon, J. & Simmelhag, V. (1971). The superstition experiment: A reexamination of its implications for the principles of adaptive behavior. *Psychological Review*, *57*, 3-43.

Stich, S. (1983). *From folk psychology to cognitive science*. Cambridge: Massachusetts Institute of Technology Press.

Taylor, C. (1964). *The explanation of behavior.* London: Routledge & Kegan Paul.

Teller, P. (1984). Comments on Kim's paper. In T. Horgan (Ed.), *Spindel Conference 1983: Supervenience* (Supplement to *The Southern Journal of Philosophy*, *22*, pp. 57-61).

Terrace, H. S. (1979). *Nim: A chimpanzee who learned sign language.* New York: Washington Square Press.

Titchener, E. B. (1898). The postulates of a structural psychology. *Philosophical Review*, *7*, 449-465.

Tinbergen, E. A. & Tinbergen, N. (1973). Early childhood autism--an ethological approach. In N. Tinbergen, *The animal in its world: Explorations of an ethologist (Vol. 2, Laboratory experiments and general papers*, pp. 175-199). Cambridge: Harvard University Press.

Vaughan, M. (1989). Rule-governed behavior in behavior analysis: A theoretical and experimental history. In S. C. Hayes, (Ed.), *Rule-governed behavior: Cognition, contingencies, and instructional control* (pp. 97-118). New York: Plenum Press.

Verplanck, W. S. (1954). Burrhus F. Skinner. In W. K. Estes, S. Koch, K. MacCorquodale, P. E. Meehl, C. G. Mueller, Jr., W. N. Schoenfeld, & W. S. Verplanck, *Modern learning theory* (pp. 267-316). New York: Appleton-Century-Crofts.

Wann, T. W. (Ed.) (1964). *Behaviorism and phenomenology: Contrasting bases for modern psychology*. Chicago: University of Chicago Press.

Watson, J. B. (1913). Psychology as the behaviorist views it. *Psychological Review*, *20*, 158-177.

Wearden, J. H. (1988). Some neglected problems in the analysis of human operant behavior. In G. Davey & C. Cullen (Eds.), *Human operant conditioning andbehavior modification* (pp. 197-224). Chichester, England: John Wiley.

Wendt, R. (1949). The development of a psychological cult. *American Psychologist*, *4*, 426.

Wessells, M. G. (1981). A critique of Skinner's views on the explanatory inadequacy of cognitive theories. *Behaviorism*, *9*, 153-170.

Wessells, M. G. (1982). A critique of Skinner's views on the obstructive character of cognitive theories. *Behaviorism*, *10*, 65-84.

Williams, B. A. (1983). Revising the principle of reinforcement. *Behaviorism*, *11*, 63-88.

Williams, B. A. (1984). Stimulus control and associative learning. *Journal of the Experimental Analysis of Behavior*, *42*, 469-484.

Williams, G. C. (1966). *Adaptation and natural selection*. Princeton: Princeton University Press.

Zeiler, M. D. (1984). The sleeping giant: Reinforcement schedules. *Journal of the Experimental Analysis of Behavior*, *42*, 485-493.

Zuriff, G. E. (1985). *Behaviorism: A conceptual reconstruction*. New York: Columbia University Press.

INDEX OF NAMES

INDEX OF SUBJECTS

STUDIES IN COGNITIVE SYSTEMS

1. J. H. Fetzer (ed.): *Aspects of Artificial Intelligence.* 1988
   ISBN 1-55608-037-9; Pb 1-55608-038-7
2. J. Kulas, J.H. Fetzer and T.L. Rankin (eds.): *Philosophy, Language, and Artificial Intelligence.* Resources for Processing Natural Language. 1988
   ISBN 1-55608-073-5
3. D.J. Cole, J.H. Fetzer and T.L. Rankin (eds.): *Philosophy, Mind and Cognitive Inquiry.* Resources for Understanding Mental Processes. 1990
   ISBN 0-7923-0427-6
4. J.H. Fetzer: *Artificial Intelligence: Its Scope and Limits.* 1990
   ISBN 0-7923-0505-1; Pb 0-7923-0548-5
5. H.E. Kyburg, Jr., R.P. Loui and G.N. Carlson (eds.): *Knowledge Representation and Defeasible Reasoning.* 1990 ISBN 0-7923-0677-5
6. J.H. Fetzer (ed.): *Epistemology and Cognition.* 1991 ISBN 0-7923-0892-1
7. E.C. Way: *Knowledge Representation and Metaphor.* 1991
   ISBN 0-7923-1005-5
8. J. Dinsmore: *Partitioned Representations.* A Study in Mental Representation, Language Understanding and Linguistic Structure. 1991 ISBN 0-7923-1348-8
9. T. Horgan and J. Tienson (eds.): *Connectionism and the Philosophy of Mind.* 1991 ISBN 0-7923-1482-4
10. J.A. Michon and A. Akyürek (eds.): *Soar: A Cognitive Architecture in Perspective.* 1992 ISBN 0-7923-1660-6
11. S.C. Coval and P.G. Campbell: *Agency in Action.* The Practical Rational Agency Machine. 1992 ISBN 0-7923-1661-4
12. S. Bringsjord: *What Robots Can and Can't Be.* 1992 ISBN 0-7923-1662-2
13. B. Indurkhya: *Metaphor and Cognition.* An Interactionist Approach. 1992
    ISBN 0-7923-1687-8
14. T.R. Colburn, J.H. Fetzer and T.L. Rankin (eds.): *Program Verification.* Fundamental Issues in Computer Science. 1993 ISBN 0-7923-1965-6
15. M. Kamppinen (ed.): *Consciousness, Cognitive Schemata, and Relativism.* Multidisciplinary Explorations in Cognitive Science. 1993
    ISBN 0-7923-2275-4
16. T.L. Smith: *Behavior and its Causes.* Philosophical Foundations of Operant Psychology. 1994 ISBN 0-7923-2815-9

KLUWER ACADEMIC PUBLISHERS – DORDRECHT / BOSTON / LONDON